GOD

what Quantum Physics tells us about

IS NOT

our origins and how we should live

DEAD

AMIT GOSWAMI, PhD

HAMPTON ROADS

Cover design by Jim Warner
Cover photograph © William Attard McCarthy
Author photograph from the film *The Quantum Activist* © 2009
by Bluedot

Hampton Roads Publishing Company, Inc.
Charlottesville, VA 22906
Distributed by Red Wheel/Weiser, LLC
www.redwheelweiser.com

Library of Congress Cataloging-in-Publication data is available
upon request.

ISBN: 978-1-57174-673-3

Printed in the United States of America
QG
10 9 8 7 6 5 4 3 2

Contents

Contents

Preface

C an the question of God's existence be settled by scientific evidence? In this book, I show that it can and it already has, in God's favor. But the evidence is subtle, based on the concept of the *primacy of consciousness* from quantum physics, which remains Greek to many people, and so the message is very slow to penetrate both scientific and popular consciousness. This book attempts to accelerate the acceptance of God once again in our society and our culture.

One question needs to be settled at the outset. What is the "God" that science is rediscovering? Everyone knows that even religious people who talk the most about God cannot agree about what God is. So, is science rediscovering a Christian God, a Hindu God, a Moslem God, a Buddhist God, a Judaic God, or a God of a less popular religion? The answer is crucial.

What almost everyone doesn't know is that at the esoteric core of all the great religions, there is much agreement about the nature of God. Even at the popular level, most religions agree about three fundamental aspects of God. First, God is an agent of causation, over and above causation arising from the material world. Second, there are more subtle levels of reality than the material level. And third, there are Godlike qualities—love is a primary one—that religions teach people to aspire to as a major goal. What is the God that science is rediscovering? Suffice it to say for now that the God rediscovered by science has all three of these important aspects.

I offer two kinds of scientific evidence for God.

The first kind I label as "the quantum signatures of the divine." Quantum physics gives us such novel aspects of reality—the quantum signatures—that to understand, explain, and appreciate them, we are forced to introduce the God hypothesis. An example is quantum nonlocality: signal-less communication. Ordinary local communication is carried out via signals carrying energy. But in 1982, Alain Aspect and his collaborators verified in the laboratory the existence of communications requiring no signals. Hitherto, the belief was that such quantum signatures occur only in the submicroscopic world of matter and somehow are not important for the macro domain, or mundane level of reality. But I demonstrate that these quantum signatures also occur at this level, and that they provide indisputable evidence for the existence of God. Research groups conducting experiments with several kinds of phenomena have found such evidence in the laboratory.

The second kind of evidence involves what religions call *subtle domains of reality*. You could very easily label this kind of evidence as pertaining to impossible problems requiring impossible solutions (from the materialist point of view, that is).

An example will make this clear. Recently, there has been a lot of controversy about creationism-intelligent design theories versus evolutionism. Why so much controversy? It is because even after 150 years of Darwinism, evolutionists do not have a foolproof theory. They cannot satisfactorily explain either the fossil data, especially the fossil gaps, or why and how life appears to be so intelligently designed. This is what creates room for controversy. Honest, unprejudiced scientific appraisal of these theories and data shows what follows.

Neither Darwinism nor its later synthesis with genetics and population biology called neo-Darwinism agrees with all the experimental data.

Creationism and intelligent design theories as proposed have little scientific content, with creationism having hardly any scientific content, but there is indisputable data supporting the fundamental ideas of both evolution and intelligent design (although not Bible-based creationism).

The key here is to ask, is there an alternative to both of these approaches that agrees with all the data? My answer is yes, and I will demonstrate it in this book. But this requires the existence of a causally empowered God and a subtle body that acts as a blueprint for biological form; materialism permits neither of these. Nevertheless, impossible problems require impossible solutions!

Another example involves the processing of meaning. The philosopher John Searle and the physicist Roger Penrose have shown that computers can process only symbols, not the meaning that the symbols may represent. For generating and processing meaning, we need the mind. But then the question arises: How does mind interact with matter? The age-old dualism problem of the mind-body interaction still haunts us. This is where I show that the God hypothesis is essential to settle the mind-body interaction problem. And in this new "impossible" context, our creative ability to process new meaning gives us much tangible scientific evidence for the existence of God.

If the good news is that such evidence for God is already here, then what should we do about it? Well, first we must reformulate our sciences within the quantum God hypothesis and demonstrate its usefulness outside of quantum physics. In this book, I demonstrate that this one hypothesis solves all the hitherto unsolved mysteries of biology: the nature and origin of life, fossil gaps of evolution, why evolution proceeds from simple to complex systems, and why biological beings have feeling and mysteriously consciousness, just to mention a few. We also find that within the quantum God hypothesis, the "depth" psychology of Sigmund Freud, Carl Jung, and James Hillman, based on the unconscious, is seen as quite complementary to the "height" psychology of the humanists and transpersonalists of recent times—Carl Rogers, Roberto Assagioli, Abraham Maslow, and Ken Wilber—based on transcendence or superconsciousness. Both these psychologies are now recognized as defining paths to the realization of God in our personal lives.

There are other aspects of the quantum God hypothesis that every one of us can appreciate and even bring to fruition. This new science validates our current preoccupation with meaning, although the materialist worldview is doing its best to undermine it. Equally important is that a God-based science puts ethics and values where they belong, at the centers of our lives and societies.

We may not like some aspects of the old religions that hitherto were the only proponents of the concept of God, but we can agree that all religions gave us ethics and values (the cultivation of Godliness) for our societies. These have been undermined by the current materialist worldview, with devastating results for our politics, economics, businesses, and education. With the scientific rediscovery of God, which also emphasizes ethics and values, we gain an opportunity to revitalize modern social systems like democracy and capitalism that seem to have become bogged down with seemingly insurmountable difficulties.

The preoccupation with meaning, ethics, and values is important for the evolution of humanity. My final message is what I call *quantum activism*: combining the usual activism of changing the world with ongoing efforts to align oneself with the evolutionary movement of the whole. If the latter step requires our creativity and quantum leaps in our processing of meaning and values while engaged in the affairs of the world, so be it. At the least this will bring new meaning and value to our lives; at best, it will usher a new age of enlightenment.

I acknowledge with deeply felt gratitude all the people who have contributed to the rediscovery of God, which is the subject of this book. The names are too many to list, with one exception: my wife Uma, who is also my partner in my current spiritual practice. I also thank all the quantum activists who have worked with me in the past or are currently working with me, and also those who will undertake quantum activism in the future. Finally, I thank my editors, Bob Friedman and John Nelson, and the staff of Hampton Roads for a fine production job on this book.

Prologue

For Skeptics

Before presenting this book to you, Dear Reader, I asked myself what the reaction would be to the basic idea of this book from three types of die-hard skeptics: the materialist scientist, the Western philosopher, and last but not least, the Christian theologian? So I decided to do an exercise employing active imagination to address the skepticism of these three groups head-on.

In my imagination, I create my straw scientist. He is a white American male, complete with coat and tie loosened at the neck (to signify openness, a touch of the eminent American physicist Richard Feynman). He has an air of all-knowing nonchalance and a lighted cigar in hand, emulating the celebrated Danish physicist Niels Bohr. And of course, he has the impatient arrogant smile of the American biologist James Watson, intended to hide his forever-present insecurity. I then ask my scientist, "I am planning to present a book about the scientific evidence for God. What do you think of the idea?"

"Not much," he says, not too surprisingly. He elaborates. "Look, we've heard such claims of scientific evidence before. Take the creationists, for example. For all the noise they make, when you look closely, all their evidence is based on the negatives of our argument. They are clever, I admit. They do make many interesting points about the holes in the Darwinian evolution theory, our antidote to their so-called creation science. But we've countered by pointing out that their ideas don't constitute science because they're not verifiable." He gives me a challenging look and continues, "Look, I know you want to make a case for God by highlighting all the shortcomings of materialist science in explaining things. But that will never work."

This is not an important part of my approach. But I am curious. "But why?"

"Why?" His smile now becomes condescending. "Because, my idealistic friend, we can always address our negatives by the promise of future scientific discoveries. The answers are blowing in the wind of future science."

"I know, I know." I too can be condescending. "Didn't Karl Popper denigrate that, calling it promissory materialism?"

His cigar has gone out and he becomes busy relighting it. He takes a long puff and spews out a cloud of smoke. Now he gives me a penetrating look, like he's ready to level me with his reply. "What is God?" he asks presently.

But I am ready for him. I say with quiet confidence, "God is the agent of downward causation."

"Oh, that hackneyed answer," he pooh-poohs. "I thought you'd have something better. We eliminated that God long ago, because it's dualism. How does a nonmaterial God dish out downward causation that affects material objects? Any interaction with the material world takes an exchange of energy. But the energy of the material world alone is always conserved. No energy ever flows out to God or comes in from God. How is that possible if God is always interacting with the world?"

"You didn't let me finish . . . ."

"And you didn't let me finish," he continues. "Look here. We don't deny that you feel the presence of an almighty God in your religious rituals. But we have an explanation: God is a brain phenomenon. When you tickle certain centers of the mid-brain with your rituals, you elicit experiences of a powerful force. Downward causation makes sense to you then. OK?"

"Not OK." I can also be firm. "God is the agent of downward causation, but it doesn't have to be the dualistic God of old. Your problem is that since Galileo you've been fighting the straw God of popular Christianity, which isn't the real issue at all. The real issue is: Can your model of reality, with one material level of existence and upward causation from the base level of matter (figure 1-1, page 17), account for everything? And it can't. You have to face up to that.

"Christians of the old traditions tried to explain everything they could not understand with the general principle: God and His downward causation. It's a very limited idea. Science was developed to fight that idea and to discover better ways for understanding the data. Today, you materialist scientists are doing the same thing. With any unexplainable phenomenon, you either deny it or try to explain it away with worn-out concepts like 'God is an emergent epiphenomenon of the brain' or 'God is a useful adaptation under the Darwinian struggle for survival.' We can never verify such ideas."

"You are lecturing me," he grumbles gruffly.

"So? You lectured me." I am stern. "The God I'm talking about is *quantum consciousness*. As you know very well, in quantum physics objects aren't determined things; instead they are possibilities for God—quantum consciousness—to choose from. God's choosing transforms the quantum possibilities into actual events experienced by an observer. Surely you accept the idea that quantum consciousness is scientific."

"Yes, of course. The observer effect: quantum objects are seemingly affected by conscious observers or by consciousness." Then he smiles slyly. "New wine in an old bottle, eh? Trying to make the idea of quantum consciousness provocative by renaming it as God?"

He is not getting the point. "Look, I am quite serious. Quantum consciousness is really what our savants, the mystics, have meant by the word 'God.' I begin my exposition p roving that and also pointing out that it's an experimentally verifiable idea."

He interrupts me. "Really? Look, the observership is just an appearance, and there must be a material explanation for this appearance. It's too hasty to postulate real consciousness." He sounds a little exasperated.

"But it's logically consistent to assume so. To do otherwise gives you a paradox."

3

"Yes, but we can't let a few paradoxes get in the way of our philosophical convictions," he says slyly.

He is not getting the point. "Look, I'm quite serious. I repeat that quantum consciousness is really what our mystics have meant by the word 'God.' Let me also repeat that it is an experimentally verifiable idea."

He now hears me and his mouth falls open. "Really? How?"

"Look, ever since the physicist Pierre-Simon Laplace told Napoleon, 'I don't need that [God] hypothesis [for my theories],' you guys have been using that argument to disprove God."

"And successfully, too," my scientist interrupts.

"Yes, but now turnabout is fair play. I'll present theoretical paradoxes and experimental data to show that we do need the God hypothesis, and not only to remove logical paradoxes from our theories but also to explain much new data. Brace yourself."

My scientist looks away. I know I've gotten to him. Scientists respect resolution of paradoxes and, most of all, experimental data.

But my scientist comes around and slyly says, "Surely you don't expect us to lay aside our convictions just because of a few paradoxes. As for new data, it's a bit speculative to say that quantum physics, designed for the micro world, also works for the macro or mundane world. This is what you're implying, isn't it? I suppose next you're going to tell me that this has already been verified by objective experiments in the macro world."

I smile. "That's exactly what I'm telling you. As to the applicability of quantum physics to the macro world, surely you know about *SQUID*?"

My scientist grins. "Squid? My wife sometimes serves it for dinner. I can't say I like it very much."

I shake my head. "You know that SQUID is the acronym for *Superconducting Quantum Interference Devices*. It's too technical to delve into it here, but those experiments showed long ago that quantum physics applies all the way to the macro world, as it should. Also, the God-verifying experiments I will discuss in this book are all macro-level experiments. Some of the new data has even been replicated."

My scientist looks a little uncomfortable. "Look here. We are never going to accept what you're doing as science. You know why? Because science by definition looks for natural explanations. You are inviting something supernatural, God, into this hypothesis. It can never be science." He sounds stubborn.

"If by 'nature' you mean the space-time-matter world, then your science cannot even accommodate quantum physics. Shame on you. The Aspect experiment—photons affecting one another without signals through space and time—settled that issue once and for all."

My scientist again looks away. His cigar has very conveniently gone out again. I get up. I know I've gotten to him. Scientists respect objective experiments. One down, the materialist scientist; two to go.

In my imagination, I now create the skeptic philosopher: tall, white American male with a shaved head and looking a lot like Ken Wilber. I tell him about my book on the scientific evidence for the existence of God. I also tell him about my encounter with the skeptical scientist. He surprises me with his question. "What is science?"

I fumble with words a bit. "We have ideas about being, through our experience of the outer and the inner worlds and through our intuitions. Those constitute our philosophy of being that you philosophers call *ontology* or *metaphysics*. Next comes how we know 'being,' which you philosophers call *epistemology*, right? Scientists intuitively theorize about being, make deductions from various theoretical insights, and then subject the theories to experimental consensus verification. Science is an epistemology with two wings: theory and experiment."

I look at him for approval. He says gruffly, "Fine, fine. But what you study and discover through this science is about manifest experience, ephemeral, wouldn't you say?"

He is right. I nod in agreement.

"Then tell me, how can you use this science of temporal phenomena, space-bound phenomena, to prove the existence of what is eternal, what is beyond all phenomena, what is transcendent? Your idea is worse than those fo the medieval Christians who tried to prove God's existence through reason, because of your scientific pretentiousness. You think people will accept your idea because you cloak it in science, don't you?"

5

This fellow is arrogant, also cynical. I try to respond, but he continues in a staccato voice. "I know of your kind of scientific proof of God. You manage to do it not only by redefining God, but by even redefining materialism. You're a holist, right?"

Actually, I am not a holist—not the usual kind who thinks that the whole is greater than its parts or that novel creations can emerge from simple components but cannot be reduced to them. But his question has perked my curiosity. "So what have you got against the holists?"

He looks at me scornfully. "Look, as even Descartes understood four hundred years ago, matter is fundamentally reductionistic: the microcosm makes up the macrocosm. To suggest that matter in bulk, because of complexity, can have novel emergent features is preposterous. You think God is an emergent interconnectedness of matter, and God's downward causation is an emergent causal principle of complex matter, but this kind of idea is easily refuted." He pauses, looking at me for a reaction. I remain quiet. He continues.

"If the idea of emergent holism held water, it would show up whenever we make complex matter out of the simple. For example, when hydrogen and oxygen mingle together to make a water molecule, does any property emerge that cannot be predicted from the interaction of the constituents? No. And if you say that the wetness of water, which we can feel, is such an emergent property, I'll hit you. Our feeling of wetness of water comes from *our* interaction with the water molecule."

I try to mollify him. "I'm not saying that anything new and holistic emerges when hydrogen and oxygen combine to make water. Actually, I agree with you. The holists walk on very thin ice."

He does not seem to hear what I said and continues, "If God were only an emergent interconnectedness of matter, God would be time-bound and space-bound, limited. There would be no transcendence, no sudden enlightenment, and no spiritual transformation. You can call the holist view deep ecology, garb it with the fancy names that satisfy mediocre minds, but it does not satisfy the philosophically astute. It does not satisfy me."

Again his arrogance is showing. And in this case, he is right, of course, on his basic point. I try to be patient and exclaim, "O great

6

philosopher, you are right. Holism is a hopeless approach of the fence-straddling philosopher who values God but won't give up materialism entirely. And you are right that science can never find answers about the ultimate truth. Truth is.

"But behold, please. Materialists make the ontological assertion that matter is the reductionistic ground of all being: everything, even consciousness, can be reduced to material building blocks, the elementary particles and their interactions. They hold that consciousness is an epiphenomenon, a secondary phenomenon of matter that is the primary reality. What I demonstrate is the necessity of turning the materialist science upside down. Quantum physics demands that science be based on the primacy of consciousness. Consciousness is the ground of all being, a being that mystics call Godhead. Let materialists realize that it is matter that is the epiphenomenon, not consciousness."

"I see." My philosopher is unruffled. "That all sounds very noble. But now haven't you gone too far the other way? Can you call it science if you base it on the primacy of consciousness?

"The way I see it, scientists can look at the objective side of consciousness, the It and Its—the third-person aspect of consciousness, so to speak. The mystics, indeed all of us, personally look at the subjective side—the first-person experience. The philosopher can do even better by considering the intersubjective side—the second-person relationship aspect. This is what I call the 1-2-3, the first person, second person, and third person aspects of consciousness. If we extend consciousness study from the purely scientific objective to include the other aspects as well, we get a complete model, the four-quadrant model (figure 3-1, page 45). The problem of consciousness is solved. We don't need quantum physics and your new paradigm thinking about science."

I am a little startled by his claims. This fellow is tough in his own way. Nevertheless, I manage to say, "That's real good. It describes the phenomenon as phenomenology; this is impeccable. But the model does not integrate the four quadrants."

He smugly retorts, "That is precisely my point and that of the mystic. To integrate, you have to go beyond science, beyond reason, into higher states of consciousness."

Now it is my turn to be tough. "This is an elitist position and you know it. Mystics have always said that in order to know reality it takes higher states of consciousness. And then they say to whoever listens, 'Be good. Because I have experienced these higher states and I know what is good for you.' But has this ploy ever worked?

"It works to some extent because being good is part of our nature; hence the appeal of religions. But base emotions are also part of our nature; hence materialism also appeals to us. And this mysticism-materialism debate goes on, in public and in private."

"So what are you proposing?"

"Quantum physics enables us to develop a dynamic integration of spiritual metaphysics and the science of the material world. It retains the mystery of mysticism, of the ultimate reality. But it allows reason to penetrate deep enough to understand the integrity of your 1-2-3 of consciousness," I say gravely.

The philosopher is now respectful. "How does that quantum redefinition of science help establish God so scientists and everyone else will accept the idea and try to be good?" he asks.

"Remember my dialogue with the scientist?" I can feel I have his full attention now. "God is quantum consciousness; this is a level below the absolute level of consciousness as the ground of all being. Scientific objectives and experimental tests can be engaged at this level—not to test God directly, but to test God's power of downward causation that manifests not only the material world but also the subtle levels. We are also finding solid objective data for the existence of the subtle. It is this objective experimental verification that will convince everyone and lead to a paradigm shift. Surely you agree?"

"All right, all right. It will certainly be interesting to read what you've got," he says with an air of dismissal. He needs to have the last word. Recognizing his need, I take my leave.

Two down and one more to go: the Christian theologian. I try to create him carefully, proper garb and everything. To my surprise, this one ends up as a woman. The world is changing indeed; there is hope for God yet.

I greet my theologian. I tell her about the title of my book and also about my bouts with the skeptical scientist and the philosopher. She chuckles quite sympathetically. Then suddenly her smile disappears as she speaks in rapid staccato.

"You know I'm sympathetic to your cause, but my skepticism comes from our experience with the materialists. Don't underestimate them; they will eat you alive."

"They sure ate you alive." I can't resist the jibe. "But you know why, don't you? You don't take science seriously, materialist though it may be so far. It took the Pope four hundred years to acknowledge Galileo and a decade longer to acknowledge Darwin. And the fundamentalists of your flock still fight the idea of evolution tooth and nail. But we take materialists seriously and respectfully; we give them their due. The new science is inclusive of materialist science."

"Fine, fine," says my theologian. "But your inclusion of their science won't please them, you know. They want to be exclusive.

"So many times we've tried to corner them arguing about the gaps in their science, trying to prove the existence of God and downward causation in those gaps. But materialists have always been able to thwart our efforts and narrow the gaps."

"We have deeper evidence than gap theology."

She interrupts me in midsentence. "I know, I know. We have deeper evidence, too. Such beautiful evidence, such beautiful arguments starting from William Paley to the current intelligent design theorists. If purposiveness is not a signature of the divine, what is? If you see a beautiful watch in a forest, how can you not see purpose, how can you ignore the designer, the watchmaker? Likewise, how can you see the beautiful living creatures of nature and not wonder about God's purpose, about God, the designer Himself?

"But the philosopher Herbert Spencer and more recently the biologist Richard Dawkins turn the intelligent design arguments around! The purposiveness of the biological world is appearance, they say. Not a signature of teleology, but mere teleonomy, its purposefulness the result of Darwinian adaptation. Dawkins even wrote a book calling God *The Blind Watchmaker*. And another one called *The God Delusion*, as if calling God a delusion will make it so. And people buy into his ideas, too. Even judges."

Actually, the last assertion is not quite true. Although a federal judge in 2006 ruled against the teaching of intelligent design in schools, that was because the case for intelligent design is somewhat weak as of yet. One of my aims for this book is to correct that.

The fact is, many scientists have seen the weakness of Dawkins' arguments through probability calculations that show the improbability of life originating from matter driven by blind chance and survival-necessity, as Dawkins pretends. But this discussion would take us sideways. I try to get back to the main point.

"Your main problem is that the picture of God you portray is so naïve that it's easy to pick apart, and Dawkins and other materialists have had a heyday doing so. They always use the God of popular Christianity as a straw God to make their point. Let them use the esoteric notions of God and see if they can disprove God using materialist arguments!

"But I am proposing more than that. Let's talk about signatures of the divine. You'll be happy to know we have a new foolproof track to finding these signatures."

"How so?" I have managed to pierce through my theologian's cynicism. Now she is openly curious.

"You see, madam, you theologians see signatures of the divine in the gaps of scientific understanding. And it is not a bad idea, per se. I respect you for it. But you have failed to discriminate between gaps that are, at least in principle, possible to bridge via the materialist approach to science and those that are unbridgeable using this approach. You have been a little wishy-washy."

"Maybe so. But what is your alternative?"

"We discriminate. We home in on those gaps that are impossible to bridge through a materialist approach. I call these the 'impossible questions for materialism.' And there is more.

"The application of quantum physics gives us another kind of signature of the divine: quantum consciousness. An example is the discontinuous insight of the creative experience, a discontinuity that today we identify as a quantum leap of thought. There are other signatures: non-local interconnectedness that operates without signals through space-time.

"These quantum signatures are made of indelible ink; they cannot be erased or rationalized away by any materialist hocus-pocus."

"Really? That is incredibly hopeful. But I have to ask you, how does your new approach regard Jesus? Does it recognize the specialness of Jesus?"

"Of course. Jesus is very special. One of a very special category of people, the perfected beings."

My theologian becomes thoughtful. "You don't subscribe to the idea that Jesus is the only begotten Son of God?"

"No. But I do the next best thing. I show that the category of people to which Jesus belonged all have regular access to a state of consciousness—call it the Holy Spirit—that is truly the only begotten Son of God."

"This is interesting. Reminds me of some new-paradigm thinking within Christian theology itself."

"That it does."

Here is the book. It is about God—quantum consciousness—a new paradigm of science based on the primacy of consciousness, and about scientifically verifiable quantum signatures of the divine that cannot be rationalized away. It is about the meaning and purpose of our spiritual journeys, and the meaning and purpose of evolution.

For millennia, we humans have intuited God and have searched. What we have found has inspired us to be good, nonviolent, and loving. But we have mostly failed to live up to our intuitions of how to be good, how to love. In our frustration, we have become defensive; we have become believers of God who have to defend the idea of God as an excuse for the inability to live up to that idea. This has given us religious proselytizing, fundamentalism, even terrorism—all in the name of God.

Modern science grew out of the effort to free ourselves from the tyranny of religious terrorism. Truth, of course, is Truth, so it is inevitable that science now has rediscovered God. Unfortunately, I doubt if this alone will make the difficulties of living the ideals of God much easier.

So are we in danger once again of creating a dogma that we have to defend out of the guilt of not being able to live up to its demands? I hope not.

One advantage of Godless materialist science is that it is value-neutral to some extent, and nobody has to live up to any ideals. In fact, it encourages people to become cynical existentialists and indulge in consumerism, maybe downright hedonism. Of course, this also creates the vast wasteland of unfulfilled human potential that we see all around us today.

The new science within consciousness comes with more understanding of where past religions, the past upholders of the concept of God, have failed. The quantum signatures of the divine tell us quite unambiguously what we need to do to realize God in our lives, why we fail, and why we hide our failure and become fundamentalist activists. If you heed the quantum signatures of the divine, the importance of quantum leaps and nonlocal knowing, you have another choice. I call this choice *quantum activism*.

Ordinary activism is based on the idea of changing the world so that you don't have to change. By contrast, spiritual teachers tell us constantly that we should concentrate on our own transformation and leave the world alone. Quantum activism invites you to take a middle path. You acknowledge the importance of your own transformation, and you travel the transformational path earnestly, the difficulties of quantum leaping and nonlocal exploration notwithstanding; but you don't say that it is transformation or bust. You also pay attention to the holomovement of consciousness that is evolving in the world around you and help it along.

So finally, the book is also an introduction to quantum activism. Needless to mention, I am a quantum activist myself. So, dear reader, welcome to my world!

Part One

Introduction

I n 1973, after about ten years of being a regular academic scientist, I was unhappy, but I did not know why. The following incident made me realize why.

I was at a nuclear physics conference; nuclear physics was the chosen field of research that engaged my heart and soul—or so I thought. I was a speaker at the conference and, when my turn came, I gave what I thought was a good presentation. Nevertheless, I was dissatisfied because I found myself comparing mine with other presentations and feeling jealous. The jealousy continued throughout the day.

In the evening I was at a party; there was lots of free food and booze along with a lot of interesting company, people to impress, etc. But I felt more of the same jealousy. Why were people not paying attention to me— not enough to relieve my jealous feelings, anyway?? This went on until I realized that I had a heartburn that wouldn't quit. I had already finished an entire packet of Tums that I carried in my pocket.

Feeling desperate, I went outside. The conference was taking place at Asilomar Conference Grounds on Monterey Bay in California. Nobody else was outside, it was a bit chilly. Suddenly, a blast of cool sea breeze hit my face. A thought surfaced (where did it come from?): "Why do I live this way?"

Why did I live this way? Paradigm research in practically every field of science consists of a few people defining the problems that require attention and others following their lead and carrying out the details. To belong to that elite group of trendsetters depends on a lot of things. The easy way for an academic is to be a follower and to publish rather than "perish" in the attempt to become a trendsetter. That was what I was doing; I was following with gusto.

Why did I live this way? Most problems of paradigm science are irrelevant to our lives. They are almost as esoteric as the questions that Christian monks in medieval times studied: how many angels can dance on a pin? So my life and my work were completely out of sync.

Why did I live this way? Is physics at all relevant to us today? Nuclear physics is relevant to weapons research, maybe energy research as well, but it is not relevant to much else. In Einstein's time, physics was relevant; in Niels Bohr's time, yes, certainly. Those were times of a paradigm shift that affected not only all of science, but the way we see the world in general.

Why did I live this way? I had academic tenure. There was no reason for me to do unhappy physics. I would find some "happy" physics to do and see.

I had no idea that the decision to pursue my personal happiness in physics would lead to a scientific rediscovery of God. I was a staunch materialist, you see.

Chapter 1

The Scientific Rediscovery of God

The concept of a higher power, popularly called God, is millennia old. The idea is that we experience phenomena that cannot be explained on the basis of material, worldly causes alone; the only explanation possible is that the phenomena are caused by intervention from God. This divine intervention is called *downward causation*.

This concept conjures up an image of God as a mighty emperor sitting on a throne up in heaven and doling out acts of downward causation: acts of creation, different laws of movement for heavenly and earthly bodies, miracle healings for devotees, judgment of the virtuous and the sinners, and so forth. Support for this naïve, outdated picture is implicit in pop religions even today, especially popular Christianity.

Scientists take advantage of the naïveté of the populist God supporters to pooh-pooh this description as dualism that is philosophically untenable, impossible. God is dishing out downward causation, intervening in our world now and then, here and there? Hah! That's impossible, they assert. How does a nonmaterial God interact with things in a material world? Two entities of different kinds cannot interact without a *mediator signal*. But the exchange of a signal involves energy. Alas! The energy of the physical world alone is always conserved or is a constant. But that would be impossible if the world were involved in any interaction with an otherworldly God! Case closed.

The populists of Christianity strike back against this argument of science with attacks on one of the most vulnerable theories of materialist science—the theory of evolution called *(neo-) Darwinism*. But these populists, known as creationists and intelligent design theorists, do not deliver any credible alternative to neo-Darwinism, let alone to dualism.

Serious proponents of the God hypothesis respond to the criticism of dualism by stating that God is everything there is, that God is both otherworldly ("transcendent") and worldly ("immanent"). This philosophy is called *monistic idealism or perennial philosophy*. Here "transcendent" means being outside this world but able to affect what is inside this world. Downward causation is exerted by a transcendent God.

But scientists, equally seriously, have questioned this sophisticated concept, disputing this definition of transcendence. How can something be otherworldly and yet be the cause of anything in this world? This concept also smacks of dualism, they insist.

Scientists long ago attempted to show that the phenomena of the world can be understood without the God hypothesis. René Descartes intuited the idea of a clockwork universe in which a supreme being caused the universe to exist as a system of bodies in motion, providing a fixed and constant amount of motion according to the laws of physics, mechanics, and geometry, and then did not subsequently intervene in any way. Galileo Galilei discovered the two-pronged approach of theory and experiment that we call science. Isaac Newton discovered the laws of physics behind the clockwork deterministic universe, laws that apply to heavenly and earthly bodies alike. Then Charles Darwin discovered an evolutionary alternative to Biblical ideas of life's creation that fits the fossil data to some extent.

These and other phenomenal successes of a Godless science have prompted the following hypothesis: *All things consist of elementary particles of matter and their interactions.* Everything in the world can be understood from this one hypothesis. Elementary particles form con-glomerates called atoms. Atoms form bigger conglomerates called mol-ecules. Molecules form cells; some of these cells (the neurons) form the conglomerate we call the brain. And the brain comes up with our ideas. These ideas include God, an idea that may be due to the arousal of a spot in the midbrain. In this philosophy called *scientific materialism* or *material monism* or simply *materialism*, cause rises upward from the ele-mentary particles. All causes are due to "upward causation" producing all effects, including our God experiences (figure 1-1).

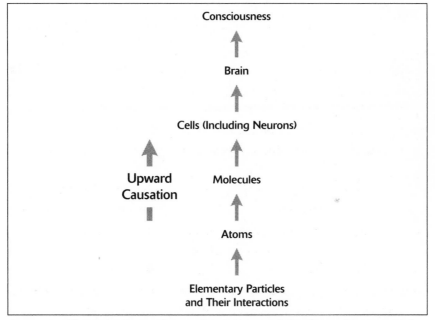

FIGURE 1-1. The upward causation model of the materialist. Cause rises upward from the elementary particles, to atoms to molecules, and so on to the more com-plex conglomerates that include the brain. In this view, consciousness is a brain phe-nomenon whose causal efficacy comes solely from the elementary particles—the base level of matter.

But the esoteric spiritual traditions say that God is beyond the brain. God is the source of our essence, the higher consciousness or Spirit in us. The question is: Does the upward causation model really explain us and our consciousness, including higher consciousness?

IS CONSCIOUSNESS A HARD QUESTION?

Currently, some philosophers have begun to call consciousness "the hard question" of science (Chalmers, 1995). Of course, such a designation depends on the context one chooses.

One context is neurophysiology, brain science, which considers that the brain generates all of our subjective experiences. Neurophysiologists posit that consciousness is an illusory ornamental epiphenomenon (secondary phenomenon) of the complex material box that we call the brain. In other words, just as the liver secretes bile, so the brain secretes consciousness.

This reminds me of a Zen story. A man meets a family of four (parents and two grown children), all of whom are enlightened. This is his opportunity to find out if enlightenment is hard or easy to attain. So he asks the father, who replies, "Enlightenment is very tough." He asks the mother, who replies, "Enlightenment is very easy." He asks the son, who replies, "It is neither difficult nor easy." Finally, he asks the daughter, who says, "Enlightenment is easy if you make it easy; it is difficult if you make it difficult."

If you think of consciousness as an epiphenomenon (secondary effect) of the brain, consciousness is a hard question indeed; you are making it hard. Consider that an objective model always seeks an answer to the question in terms of objects. Thus neurophysiologists seek to understand consciousness in terms of other objects: brain, neurons, etc. The underlying assumption is that consciousness is an object. But consciousness is also a subject—that which does the looking at and thinking about object(s). This subject-aspect of consciousness exposes one weakness of the neurophysiological brain-based model.

The truth is that consciousness is not only a hard question, but also an impossible question for materialists. This is because even pop religions, simplistic as their view of downward causation may be, have always been clear about one thing: that we have free will, and that without our free will to choose God, His power of downward causation would be in vain. If we are choosing God, defined as the highest good, we are choosing values and ethics. But we need free will to be able to make that choice.

But if we have *free* will, there must be a source of causality out-side of the material universe. So the proponents of upward causation vigorously dispute the concept of free will. If we have free will, then the behaviorist's depiction of us as the products of psychosocial con-ditioning does not work so well. They challenge the concept. Like our consciousness, our free will must also be an illusory epiphenomenon of the brain. Insisting that we are behaviorally determined machines or walking zombies, their science not only undermines God and reli-gion but also values and ethics, the very foundations of our societies and cultures.

So is there God and downward causation? Is consciousness an epiphenomenon of matter? Do we have free will? Is the dictum of the upward causation model final? Or is there new scientific evidence to suggest otherwise?

Yes, there is evidence. A revolution in physics took place at the beginning of the last century with the discoveries of quantum physics. The message of quantum physics is: Yes, there is a God. You can call it *quantum consciousness,* if you like. Some people call it by a more objec-tive phrase, *quantum vacuum field,* or following Eastern wisdom, *akashic field* (Laszlo, 2004). But a rose by any other name retains its fragrance.

QUANTUM PHYSICS: THE BASICS

The essence of quantum physics is difficult for scientists to understand; but in my experience, nonscientists have an easier time comprehending it. There are books that explain the scientists' difficulty at length. Here we can present only a quick overview.

Quantum physics is a physical science that was discovered to explain the nature and behavior of matter and energy on the scale of atoms and subatomic particles, but now is believed to hold for all matter. Scientists can describe subatomic particles only in terms of how they interact. That's how the quantum theory started, as a way to explain the mechanics of very small things. But quantum physics is now also the basis for our understanding of very large objects, such as stars and galaxies, and cosmological events, such as the Big Bang.

The foundations of quantum physics date from the early 1800s. However, what we know as quantum physics started with the work of Max Planck in 1900. The mathematics of quantum physics was discovered by Werner Heisenberg and Erwin Schrödinger in the mid 1920s.

In his quantum theory, Planck hypothesized that energy exists in units in the same way as matter, not as a constant electromagnetic wave, as had been formerly believed. He postulated that energy is *quantized*—consisting of discrete units. The existence of these units—Planck named the unit *quantum*—became the first great discovery of quantum theory.

Central to the theory of quantum physics is that all matter exhibits the properties of both *particles* (localized objects such as tiny pellets) and *waves* (disturbances or variations that propagate progressively from point to point). This central concept, that particles and waves are two aspects of a material object, is called *wave-particle duality*. It is also universally agreed that waves of quantum objects are waves of possibility.

Various interpretations have been proposed to explain this duality and other subtleties of quantum physics. One that dominated for years is known as the Copenhagen interpretation of quantum theory. This term actually refers to several interpretations, some quite at odds.

The Copenhagen interpretation is usually understood as stating that every quantum object is described by its *wave function*, which is a mathematical function used to determine the probability for that object to be found in any location when it is measured.

Each measurement causes a change in the state of matter from a wave of possibility to a particle of actuality. This change is known as the *collapse of the wave function*. In simple terms, this is the reduction of all the possibilities of the wave aspect into one temporary certainty of the particle aspect.

Unfortunately, neither the quantum mathematics nor the Copenhagen interpretation can give a satisfactory explanation of the event of collapse. But quantum physicists have been unable to eliminate the concept of collapse from the theory. The truth is, an understanding of collapse requires consciousness (von Neumann, 1955). If we follow this thinking, it means that without consciousness there is no collapse, no material particles, no materiality.

OK, so there are the bare basics of quantum physics. Now, back to the application.

QUANTUM PHYSICS AND CONSCIOUSNESS

To be sure, the mathematics of quantum physics is deterministic and based on the upward causation model. Yet it predicts objects and their movements not as determined events (as in Newtonian physics) but as *possibilities—waves of possibility* mathematically described by this wave function as mentioned above. The probabilities for these possibilities can be calculated with quantum mathematics, enabling us to develop a very successful predictive science for a large number of objects and/or events. This is the part of quantum physics that does not embarrass materialists.

Unfortunately, there is a very embarrassing aspect to quantum physics—the collapse event: a proper understanding of it revives God within science. When we look at a quantum object, we don't experience it as a bundle of possibilities, but as an actual localized event, much like a Newtonian particle. And yet, as mentioned above, quantum physics does not have any mechanism or mathematics to explain this "collapse" of possibilities into a single event of actual manifest experience. In fact, quantum physics flatly declares that there is a limit to the mathematics-based certainty of physics. There cannot be any mathematics that would allow us to connect the deterministic quantum possibilities with the actuality of a single observed event. So then, how do the quantum possibilities become an actuality of experience simply through the interaction of our consciousness, by simply us observing them (figure 1-2)? How do we explain this mysterious "observer effect"?

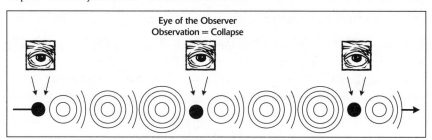

FIGURE 1-2. Quantum possibility waves and downward causation as conscious choice producing collapse.

In quantum language, the neurophysiologists' upward causation model translates like this: possible movements of elementary particles make up possible movements of atoms, which make up possible movements of molecules, which make up possible movements of cells, which make up possible brain states and make up consciousness. Consciousness itself, then, is a conglomerate of possibilities; call it a *wave of possibility*. How can a wave of possibility collapse another wave of possibility by interacting with it? If you couple possibility with possibility, all you get is a bigger possibility, not an actuality.

Suppose you imagine a possible influx of money in your bank account. Couple that with all the possible cars that you can imagine. Will this exercise ever actualize a car in your garage?

Face it. In the neurophysiological epiphenomenal model of consciousness, the assertion that our looking at something can change possibility into actuality is a logical paradox. And a paradox is a reliable indicator that the neurophysiological model of our consciousness is faulty or incomplete at best.

The paradox remains until you recognize two things. First, that quantum possibilities are possibilities of consciousness itself, which is the ground of all being. This takes us back to the philosophy of monistic idealism. Second, that our looking is tantamount to choosing, from among all the quantum possibilities, the one unique facet that becomes our experienced actuality.

To clarify the situation, let's examine how gestalt pictures are perceived—what appears at first to be one picture is actually two pictures. You may have seen the one that depicts both a young woman and an old woman, which the artist calls "My Wife and My Mother-in-Law." Another one depicts both a vase and two faces (figure 1-3). You notice that you are not affecting the picture when you shift from one perception to the other. Both possibilities are already within you. You are just making a choice between them by choosing your perspective. In this way, a transcendent consciousness can exert downward causation without dualism.

The strict materialist can still object: how can reality be so subjective that each of us observers can choose our own realities from quantum possibilities? How can there be any consensus reality in that case? Without consensus reality, how can there be science?

Surprise, surprise. We don't choose in our ordinary state of individual consciousness that we call the ego, the subjective aspect of ourselves that the behaviorist studies and that is the result of conditioning. Instead, we choose from an unconditioned, objective state of unitive consciousness, the non-ordinary state where we are one, a state we can readily identify with God (Bass, 1971; Goswami, 1989, 1993; Blood, 1993, 2001; also see chapter 5).

FIGURE 1-3 The vase and two faces. You don't have to do anything to the picture to choose either meaning.

THE QUANTUM SIGNATURES OF GOD

Here, then, are the crucial points that are worth repeating. We experience a quantum object, but only when we choose a particular facet of its possibility wave; only then, the quantum possibilities of an object transform into an actual event of our experience. And in the state from which we choose, we are all one: we are in God-consciousness. Our exercise of choice, the event quantum physicists call the *collapse* of the quantum possibility wave, is God's exercise of the power of downward causation. And the way God's downward causation works is this: for many objects and many events, the choice is made in such a way that objective predictions of quantum probability hold; yet, in individual events, the scope of creative subjectivity is retained.

In this way, the first and foremost scientific evidence for the existence of God is the vast array of evidence that supports the validity of quantum physics (which hardly anybody doubts) and the validity of our particular interpretation of quantum physics (for which there are some doubters).

Fortunately, there are two scientific ways to resolve these doubts: first, by demonstrating that this interpretation resolves logical paradoxes (rather than raising them, as does the upward causation model), and second, by making predictions that can be experimentally verified. The scientific evidence for the existence of God, based on the primacy of consciousness (the theory that consciousness creates reality) and the interpretation of quantum physics that I am presenting, passes both these tests of scientific validity. For future reference, we call this *science within consciousness* (a term first proposed by philosopher Willis Harman) or simply *idealist science*.

Phenomena resulting from downward causation in our model sometimes come with specific quantum signatures that upward causation cannot generate. If caused by upward causation—that is, if possible movements of elementary particles cause a linear hierarchy of increasing complexity that results in our consciousness—macroscopic phenomena of the mundane world would always be continuous, always consist of local communications with clear signals, and always be hierarchical in one way. The quantum signatures of downward causation are discontinuity (as in our experience of creative insight), nonlocality (as in the signal-less communication of mental telepathy), and circular hierarchy, also called tangled hierarchy (as sometimes experienced between people in love). This first kind of evidence for the existence of God I call the *quantum signatures of the divine*. The details will come later (see chapter 5); here I give you a sneak preview of one of these signatures.

It was Werner Heisenberg, one of the founders of quantum physics, who first unambiguously stated that quantum possibilities reside in transcendent *potentia*, a domain outside space and time. Quantum collapse, downward causation (the effect of our consciousness), must then be nonlocal: something outside space and time is affecting an event inside space and time. And then Alain Aspect, Jean Dalibar, and Gérard Roger (1982) brought quantum nonlocality (which implies that causes and effects can occur at a distance without an exchange of energy signals) to the experimental arena by demonstrating nonlocal connection between correlated photons (discrete objects called *quanta* of light) across a distance in a laboratory. Later measurement increased the distance of nonlocal communication between the correlated photons to more than a kilometer. Quantum nonlocality is for real.

Two things to bear in mind. First, it has become a bad habit of scientists to claim that science is about finding a "natural" explanation for phenomena while defining "nature" as the space-time-matter world. In this view, God and the subtle worlds of spiritual traditions belong to "supernature." In view of quantum nonlocality, clearly we must broaden this narrow view of nature. If science is to include quantum physics, then nature must include the transcendent domain of quantum potentia, the resident address of all quantum possibilities. In the view of quantum physics, all attempts to distinguish between nature and "supernature" have lost complete credibility.

Second, quantum nonlocality completely clarifies one confusing component of the esoteric spiritual model of God, that God is both transcendent and immanent: how some cause outside can affect something inside. This can happen because both the cause and the effect involve quantum nonlocality—signal-less interaction or communication.

A SECOND KIND OF EVIDENCE: IMPOSSIBLE PROBLEMS REQUIRE IMPOSSIBLE SOLUTIONS

Materialist science has had much spectacular success and has given us many useful technologies, but the more we apply it to biological and human problems, the less it seems capable of giving us palpable solutions. One key to developing a science with real solutions for human problems is to realize that what we experience as matter is but one important domain of the many domains of quantum possibilities of consciousness—the domain that we experience through our senses.

The psychologist Carl Jung discovered empirically that there are three more domains of conscious possibilities that we experience: feeling (of vitality), thinking (of meaning), and intuition (of supramental themes—archetypes—that we value) (figure 1-4). Recent work by Rupert Sheldrake (1981), Roger Penrose (1989), and the author

25

(Goswami, 1999, 2001) has established that feeling, thinking, and intuition, respectively, cannot be reduced to material movement; they really do belong to independent domains or compartments of consciousness. These domains are variously recognized as the *vital energy body* that we feel, the *mental meaning body* that we think, and the *supramental theme body of consciousness* (archetypes) that we intuit. All these compartments are nonlocally connected (without signals) through consciousness; consciousness mediates their interaction and there is no dualism involved (figure 1-5).

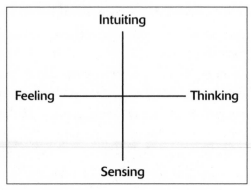

FIGURE 1-4. The four ways of experiencing according to Jung. The dominance of one or another gives us four personality traits.

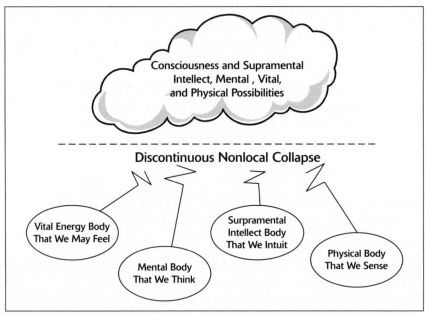

FIGURE 1-5. Quantum psychophysical parallelism. Consciousness mediates for physical, vital, mental, and supramental domains of quantum possibilities functioning in parallel.

Try to comprehend this figure; this is a breakthrough in the logjam of our thinking that has existed ever since Descartes. Our "inner" psyche (the conglomerate of vital, mental, and supramental that we experience as inner) and "outer" material world are not separate; they are parallel, ongoing possibilities of one interconnectedness that we call consciousness. This way of conceptualizing can be called a *quantum psychophysical parallelism*. It is consciousness that maintains the parallelism of the inner psyche and the outer world, and it is consciousness that causally chooses the experiences of both the outer and the parallel inner thus mediating between them. In the process, consciousness projects representations of the "subtle" inner onto the "gross" outer to experience the subtle in gross manifestation. It is like drawing a sketch of a subtle mental picture on a gross canvas to see it better. The mental picture acts like a blueprint that you represent on canvas. (How does the outer-inner distinction arise? This is explained in chapter 10.)

This is the central secret of how the world operates. Manifest reality, the world of our inner and outer experiences, is run by one central intentionality: to allow quantum consciousness, God, to experience its subtlest aspects, the supramental archetypes (such as love) in gross manifestation. So far in its evolution, consciousness has been using blueprints—the vital and the mental—to make the manifest representations (software) of the supramental on the physical (hardware). The future of our evolution can now also be told: consciousness some day will additionally make direct representations of the archetypes onto the physical, and heaven will descend to the earth, so to speak.

If you are tuned to the religious and the spiritual, here you will hear the echo of the Biblical saying, "God makes us in His/Her own image." At first, when you don't understand it, the statement jars you. Can Adolf Hitler be God's image? "Image" means representation. So far in our evolution, the representational process, image making, has been less than perfect. God has been using the blueprints of the vital and the mental. And the results have been rough and progress slow. But the prognosis for the future is glorious.

You can also understand something else. The reason that the material compartment has historically dominated our science is that the material makes (quasi) permanent representations of the experiences of the subtle levels of the psyche. Once the representations (software) are made in the material (hardware), we tend to forget the maker (consciousness) and the representation-making process (of using the blueprints—mind and the vital body).

Basically, what then is emerging is a second kind of scientific evidence for God. This consists of recognizing the many domains or "mansions" in which God's downward causation takes abode beyond the material mansion (as for example, feeling, thinking, and intuiting). Phenomena in these nonmaterial domains are all impossible problems for the materialist's upward causation model. And hence they require the solution that's impossible from a materialist's point of view: downward causation from God. Naturally, the introduction of these ideas is revolutionizing biology, psychology, and medicine. (See Parts Two, Three, and Four.)

CAMOUFLAGE

It is our patterns of habit, the ego/character that is the locus of our psychosocial conditioning, that camouflage God and the oneness of quantum consciousness. Why is this camouflage necessary? The answer is important. Our egos are necessary to give us a reference point. Without the ego, who would we be?

Similarly, the material macro world of massive objects acts as a camouflage that hides their quantum nature. Like all waves, quantum possibility waves also spread. When an electron is released at rest in a room, its wave of possibility spreads so fast that in a few moments it fills the room (in possibility): it is possible to detect the electron in various places in the room with varying probability. But in quantum mathematics, massive objects expand very sluggishly as a wave of possibility. Yet expand they do, make no mistake about it. To see through the camouflage, you must not get sidetracked by trying to see any runaway movement of the micro components of a macro body, which are bound to the center of their mass. They do their quantum waving while standing in place. Really, in the time it takes you to blink your eyes, the center of a macro object's mass is able to move by one million-trillionth of a centimeter or so. This movement is imperceptible to our eyes, but physicists, with their wonderful laser instruments, have measured such quantum movements.

Why such a camouflage? Again, it is to give us reference points for our physical bodies. If you and I manifest some of the same stuff at basically the same places every time we look, we can talk about it with one another; we can build a consensus reality. This is important. Even more important, macro physical objects can be used to represent subtler quantum objects, such as thoughts, that do tend to run away when we are not observing them. It is a good thing, too. Imagine how you would feel if, as you were reading this page, the printed letters were running away before your eyes due to their quantum movements. Of course, there is a downside to this fixity—we develop the misconception that the world of macro objects is separate from us!

To discover that we are not separate from the universe, that the entire world is our playground, we have to penetrate both of these camouflages. We have to move beyond the ego-conditioning. We have to stop being so enamored of the macro physical outer environment and look at the subtle inner environment, where objects move about with their quantum freedom much more intact.

The sun rises in the East and sets in the West. Our ancestors understood this as the evidence that the sun moves around the earth. Today we see it differently, as the evidence that the earth moves around its own axis. This explanation allows for further expansion of our understanding—that the earth moves around the sun rather than the sun around the earth. Similarly, the macro physical world has certain fixities. You can understand this through Newtonian physics and conclude that there is a world out there. Or you can discern that, because the possibility waves of macro objects are sluggish to expand, it is creating the impression that there is a world out there. In other words, there is no such world until you look! This, too, will open enormous doorways for your understanding.

If you learn to think the quantum way, it expands your mind; maybe the movement of thought is also quantum movement. You may ask, is there a way to ascertain the quantum nature of thought without going beyond conditioning? Yes. When you follow the direction of your thought, as when you free-associate during creative thinking, have you noticed how you lose the content of your thoughts? Similarly, if you focus on content, as when you meditate on a mantra, notice that you lose track of where your thought is going. In quantum physics, we call this an *uncertainty principle*, a sophisticated signature of quantum movement. If thoughts were Newtonian movement, this kind of restriction would never arise (Bohm, 1951).

I read a book, *Precision Nirvana*, in which the author, Deane H. Shapiro, illustrated what I am trying to say with two cartoons. In the first one, a good-looking girl, wide-eyed and bushy-tailed, is asking a bearded scientist type, "Professor, how do you know so much?" To this the professor replies, looking smug, "Because I open my eyes." In the second cartoon, a student is asking a Zen master serenely sitting in closed-eye meditation, "Master, how do you know so much?" To this the Zen master says, "Because I close my eyes."

Indeed, the materialist scientists cannot get over the wonders of the outer being forever bound by its camouflage. So blinded they are by the camouflage that they even try to apply their science of the outer world to denigrate the inner as epiphenomena. Didn't Abraham Maslow say that if you have a hammer in your hand, you see every problem as a nail?

And indeed, it is the effort to penetrate this camouflage that has given us the very mature spiritual traditions and their methods for reaching subtle states of consciousness beyond the ego. The camouflage of the separateness of macro objects dissolves from such subtle states of consciousness. But can one see the unity of the outer and the inner, body and mind, without the benefit of higher consciousness?

The paradigm shift of our science now taking place is revealed in depth psychology and transpersonal psychology and the branch of medicine that is called alternative medicine. The paradigm shift is also revealed in the work of organismic biologists who see causal autonomy in the entire biological organism, not merely in its microscopic components. Some evolutionary biologists even see the necessity of invoking "intelligent design" of life to break the shackle of Darwinian beliefs. The practitioners of these branches of science have penetrated the camouflage to some extent. With the help of quantum physics, the penetration of the camouflage is much more extensive, as you will see.

Quantum physics, the visionary window to the subtle, is itself very subtle; it has to be. The Nobel laureate physicist Richard Feynman used to say, "Nobody understands quantum mechanics." But he was only talking of materialists. If you are willing to look beyond the remaining vestiges of materialist beliefs, or at least if you are ready to suspend your disbelief about the primacy of consciousness and God, you've already made more progress in understanding quantum physics than many physicists and scientists.

What the Dance Is

To summarize, the old science gave us upward causation and possibilities; the new science rediscovers the agency of choice from these possibilities: God and downward causation. Together they give us the manifest reality where freedom (of the possibility wave) seeks its home in temporary bondage (of the manifest particle).

Descartes, Galileo, and Newton get the credit for most of the old scientific ideas that began the era of what philosophers call *modernism*. One of Descartes' ideas was inner (which he called mind) and outer (matter) dualism, and we are just now overturning it, although the debate over whether the monism is one based on matter or on consciousness (or God) will probably continue for a while. Descartes also gave us the philosophy of reductionism, and it has had enormous success in the material realm. But as Descartes himself recognized (unfortunately in the context of dualism), reductionism does not describe the workings of the inner realm. There one has to remember the movement of the whole. The outer fragmentation makes us individuals; the inner holism gives us feeling, meaning, goals, and purpose. Together the individual and the whole make up the partners for the dance of reality.

The legacy of Descartes, Galileo, and Newton is causal determinism, giving the scientist the hope of total knowledge and total control over reality. But it fails even for the material realm, in the submicroscopic domain where quantum indeterminacy reigns. Even so, the lure of control and the power that comes with it is so enchanting that most scientists continue to believe in causal determinism. Downward causation, which is free and potentially unpredictable, is anathema to these scientists. God they don't mind as long as it is a benign God.

The breakdown of causal determinism is just a trickle in the realm of submicroscopic physics. This is because in the material domain, at least statistical determinism holds, God builds the material world in such a way as to give us a reference point. But the trickle of freedom becomes an avalanche when it comes to the affairs of the inner. It's important to note that creativity requires movement toward the new as well as the fixity of the old. The outer—soma—gives us the fixity and the inner—psyche—gives us new movement. Together they make the dance of reality creative.

Chapter 2

The Three Fundamentals of Religion

Religions, our primary social reminder of God's existence, have been with us for millennia, starting early in human civilization. First there were primitive religions that saw two kinds of causes of events—causes that people could control (e.g., if two stones were banged together, it could make sparks to cause dry leaves to burst into fire) and causes that seemed to be outside human control (e.g., natural disasters like earthquakes). Our primitive ancestors attributed the uncontrollable causes to the agency of the gods: *downward causation*. The initial concept of multiple agents of downward causation eventually gave way to the idea of one agent: God.

With the passage of time, religious thinking became more sophisticated. The concept of God and downward causation is still there. But there is an additional concept, no less important—the concept of the individual soul or *subtle body* (a collective term for life force, mind, and consciousness). The soul is not physical, consisting of subtle substances quite different from physical substance.

And finally came the discovery that humans should pursue virtues of *Godliness*: qualities such as kindness, charity, and justice. If they didn't, they would be committing *sin* and their souls would be punished after their death.

Development in religious thinking greatly refined the picture of God and downward causation, the nature of our subtle bodies, and the ideas of virtue and sin. But these three ideas remained fundamental in religious thinking. Today virtually all religions agree about downward causation, subtle nonmaterial bodies, and the idea of ethics and morality—the ability to distinguish between virtue and sin and to choose to be virtuous. These are the three fundamentals of religion.

I point this out before presenting the scientific data for God's existence because materialist scientists, especially the Westerners among them, invariably battle against a straw God, the "superhuman" God of popular Christianity with ideas such as creationism that are easy to refute (Dawkins, 2006). But in view of quantum physics (Goswami, 1993), the vast data on life after death (Goswami, 2001), and alternative subtle-body medicine (Goswami, 2004), it is considerably more difficult to refute the ideas of downward causation and subtle bodies. And who in their right mind would try to refute the importance of virtues and values in our lives? Clearly, the religions have a more plausible theory of virtues and values than the biologists who claim they evolved from Darwinian adaptation via chance and necessity.

But materialists do make an important point: that it is difficult to talk of God within science when religions have not yet settled the big question, "What is God?" If religions still fight among themselves about whose God is superior, how can a monolithic approach such as science apply?

One answer to this kind of opposition to studying God within science is that the great traditions of the world, the major religions, are united, at the esoteric core at least, in their philosophy of God as not being dualistic. In esoterism, there is the picture of the Godhead or consciousness (or the Great Void) as the ground of all being. Within this ground, there is the concept of disparate *subtle* (immaterial) and *gross* (material) bodies. The highest ideals of human existence—loving kindness, for example—define the soul, which we try to fulfill. When we do, we are free, we are enlightened, and our ignorance is gone (Schuon, 1984).

But esoterism by itself remains obscure. The fact is that, at the popular level, most religions teach dualism: God separate from the world. And the details of this particular dual existence remain quite different from one religion to another. Well, then, isn't the point raised by the materialists valid? Let religions agree first; only then should science consider the question of God.

MULTICULTURALISM

But these scientists have not heeded the lesson of cultural anthropology. Cultural anthropologists have been arguing for some time that the idea of monolithic science may not be useful, even correct. According to them, science should be pluralistic, dependent on different cultures. Scientists tend to reject this view because they abhor the chaos arising from different points of view presented simultaneously as explanatory principles.

I think the cultural anthropologists have a point as far as the phenomena involving subtle bodies are concerned. I also believe that multicultural science does not necessarily have to be chaotic.

By and large, there is now only one physics. For the gross material bodies, the idea of a pluralistic approach is no longer necessary. The success of the reductionists' approach to physics has resolved the question in favor of a monolithic physics. But this is certainly not true for psychology and medicine, or even for biology.

In psychology, there remain three potent forces: Alfred Adler's behavioral-cognitive psychology; depth psychology, based on the concept of the unconscious from Freudian psychoanalysis and Jungian analytical psychology; and humanistic/transpersonal psychology with its concept of the superconscious. There is much data for the validity of all the approaches. For cognitive laboratory psychology, the behavioral approach is in place and mostly works. But for psychotherapy, depth psychology is a must. And for the psychology of well-being, the humanistic/transpersonal approach has its appeal and successes. So the area of psychology is a bit chaotic. There is no proper way of defining the respective domain of each of these three forces, and no attempt within psychology has succeeded in integrating them within a coherent whole.

In medicine, there are two well-known and successful approaches: conventional allopathic medicine and the different paradigms of alternative medicine. There is much bickering and chaos and little agreement as to the validity of different domains and their respective paradigms of medicine. So are we stuck with the chaos of a pluralistic approach here?

Among biologists, although there is almost universal agreement on a paradigm whose two pivots are molecular biology and (neo-) Darwinism, nobody has been able to connect this paradigm to physics or to unequivocally distinguish between life and non-life. In particular, nobody has been able to explain away the gaps in the fossil records of evolution. Therefore, a creationist/intelligent design approach to evolution continues to have popular appeal, even with some serious biologists. There are other paradigms that are gaining strength. One is based on the importance of the whole organism; we can call this an organismic paradigm. However, nobody has made a connection between the materialist and the organismic paradigms, let alone a connection between these two approaches and the intelligent design paradigm.

I submit that these difficulties of psychology, medicine, and biology arise from the fact that in these sciences, not only must the gross material body be included, but also our subtle bodies. And our pictures of the subtle bodies have not yet been refined enough to develop a useful monolithic science. Now that we have a basis, the quantum psychophysical parallelism (figure 1-5, page 26), to treat the gross and the subtle on the same footing, we have an opportunity for a much needed integrated approach, as will be demonstrated in this book.

Here then, I think, is the answer to the question, "Why are the religions so different in their details?" Because, unlike monolithic physics, the religions don't deal with the gross aspect of reality, matter. Instead, their subject is the subtlest of the subtle: God and soul.

Materialists worry that the multiplicity of religious beliefs about God is wrongheaded. Sam Harris, in *The End of Faith: Religion, Terror, and the Future of Reason*, writes, "The very ideal of religious tolerance—born of the notion that every human being should be free to believe whatever he wants about God—is one of the principal forces driving us toward the abyss." Such concerns arise from concentrating only on the differences among the religions.

We really should not bother about these differences; instead, we should concentrate on the concerns common to all religions: the three fundamentals of downward causation, subtle bodies, and achieving godliness. Here there is commonality in the religious concepts of God, and it is this commonality that allows for a scientific approach.

NEW DATA AND OUTLOOK FOR AN INTEGRATED APPROACH

In Parts 2, 3, and 4, I discuss scientific data in favor of these three fundamentals from the broad paradigm of science within consciousness defined in chapter 1. Previously, I stated that the data are of two kinds. One consists of the "quantum signatures of the divine." The other pertains to "impossible questions needing impossible answers," or subtle bodies. In truth, in much of the actual data, these two ideas get interspersed: that is, the data pertain to the subtle bodies and also are quantum signatures of the divine.

When we include in our science the subtle bodies and the quantum thinking about them, all the controversies of biology, medicine, and psychology—the chaos created by multicultural, pluralistic thinking—give way to a new integrated scientific point of view in each field. The multiculturalism is still of some use, but the domain of each culture is clearly defined and there can be free trade between them. Isn't that better?

And this kindles a new hope in me. If the various multicultural approaches to these life sciences can be integrated under one umbrella, science within consciousness, then why not the religions? Perhaps the new God-based science explored here, with all the supportive evidence in its favor, will encourage the great world religions to begin serious dialogues with one another. Perhaps we will soon have universal notions of spirituality applicable for the whole of humanity, within which each of the current religions will be a well-defined domain of validity. And there will be unlimited trade among the religions.

In the 15th and 16th centuries, religion was the grand inquisitor and the cause of much atrocity in its attempt to suppress science. But now, in an ironic role reversal, science influenced by materialism has become a grand inquisitor, arrogantly and arbitrarily declaring God and the subtle to be supernatural and superfluous. But, as I have argued above, this position leads nowhere.

As politicians under the influence of materialist science start pushing the ancient traditions to change too fast, it has the opposite effect. Instead of making badly needed changes (for example, the equal treatment of women), members of these religions become defensive, ultra-conservative, and worse. Under materialist influence, their leaders become cynical and abandon ethics and values and opt for power. If materialist science can come to terms with its own inadequacies and accept the wider arena of science within consciousness, then a new dialogue can begin between materialism and spirituality, two gravitational forces that have divided humanity through millennia. The gentle consequences of this dialogue will bring winds of change even within the old religious traditions.

Chapter 3

A Brief History of Philosophies That Guide Human Societies

There are three important philosophical "isms" that are part of most belief systems even today: *dualism, material monism*, and *monistic idealism*.

The most popular one, dualism, is also the oldest. Dualism is empirically "obvious" in our own experience because of its internal/external dichotomy. No doubt this is the reason for its popularity. In religious thinking, dualism exists as a God/world dualism: God is separate from the world but exerting influence (downward causation) on it. This dualism has dominated humanity for millennia, especially in the West. But in 17th-century Europe, René Descartes formulated a "modern" version of mind-body dualism, with the mind being God's territory, where we have free will, and the body (physical world) being the territory of deterministic science. This Cartesian dualism—a truce between science and religion—has been very influential on subsequent Western academic philosophical thinking. It also defined the modern era of Western philosophy: modernism.

Before modernism, Western society was in the severe doldrums of the Dark Ages, when religion (in the form of Christianity) ruled unchallenged over society. Modernism freed the scientists from the grip of religion. They then set out to discover the meaning of the material world—the laws of nature—in order to gain power and control over it. And this they did with such gusto, with a technology of such unquestioned virtuosity, that their spirit pervaded all of Western society. Soon religious hierarchy and feudalism gave way to democracy and capitalism, the crowning achievements of a modernist society.

Soon after, buoyed by the success of science, people began to question the necessity for this truce between science and religion. In truth, dualism does not stand up well to such obvious questions as these: How do the two bodies made of two entirely different substances interact? How does God of divine substance interact with the material world? How does a nonmaterial mind interact with the material body?

This interaction is impossible, if we allow only local interactions that are mediated by energy-carrying signals going through space and time from one body to another. An interaction between nonmaterial and material would be a violation of physics' sacrosanct law of conservation of energy, which states that the total amount of energy in an isolated system remains constant, although it may change in form. Also, there is the thorny question about the means by which this interaction would occur. What is the mediator signal made of? We seem to need a mediator made of both substances, but none exsists!

Thus material monism arose as the alternative to dualism. In material monism, the difficulties of dualism are avoided by simply insisting that there are not two substances—there is only physical matter. So, consciousness, God, our minds, and all our internal experiences are the results of the brain's interactions. These are ultimately traceable to the interactions of elementary particles (upward causation).

This philosophy has gained much credibility recently. This is not only because of its simplicity, but also because such conglomerates of elementary particles as atomic nuclei have been verified in spectacular form (nuclear detonations).

But the success of material monism also put a damper on the modernistic spirit of the West and a postmodern malaise set in. After all, if materialism holds true, then we cannot conquer and control nature as we thought we could when modernism prevailed. Instead, we humans, like the rest of nature, are determinate machines. We do not have free will or the freedom to pursue meaning as we see fit. Instead, there is no meaning in the mechanistic universe. Under the circumstances, the best we can do is to subscribe to the philosophy of existentialism: there is no meaning to our lives—each of us as an individual creates meaning (essence) in his or her life. After all, we exist somehow. Since we cannot deny our existence, we might as well play the game as it seems to be demanded of us. We pretend that meaning exists and that love exists in an otherwise meaningless, loveless universe.

This existential and pessimistic escape to nihilism—the philosopher Friedrich Nietzsche put the message well, "God is dead"—did not last long, however. Some scientists fought back with *holism*, a new idea that came from a South African politician, Jan Smuts, in his book, *Holism and Evolution,* in 1926. It was originally defined as "the tendency in nature to form wholes that are greater than the sum of the parts through creative evolution." Many scientists refused to relinquish God and religion entirely; in holism, they saw an opportunity to recover a God of sorts.

In certain primitive, animistic thinking, God exists as an immanent God, a nature God. The idea is that nature itself is animated with God. You don't have to look for God outside of this world; God is right here. Using holistic language, this can be made into an attractive philosophy. The whole cannot be reduced to its parts. Elementary particles make atoms; but atoms are a whole and cannot be completely reduced to their parts, the elementary particles. The same thing happens when atoms make molecules; something new emerges in the whole that cannot be reduced to the atomic level of being. When molecules make the living cell, the new holistic principle that emerges can be identified as life (Maturana and Varella, 1992; Capra, 1996). When cells called neurons form the brain, the new emergent holistic principle can be identified as mind. And the totality of all life and all mind, the whole of nature itself, can be seen as God. Some people see it as Gaia, the earth mother, following the ideas of chemist James Lovelock (1982) and biologist Lynn Margulis (1993).

Concurrently, this holistic thinking gave rise to the ecology movement—the preservation of nature—and to the philosophy of deep ecol-

41

ogy (Devall and Sessions, 1985)—spiritual transformation through the love and appreciation of nature itself. But materialist scientists make the valid point that matter is fundamentally reductionistic, as myriad experiments show; therefore, holism is philosophical fancy.

But there has been another alternative to dualism since antiquity: monistic idealism. Interestingly, in Greek thinking (which most influenced Western civilization), monistic idealism (enunciated by Parmenides, Socrates, and Plato) and material monism (formulated by Democritus) are almost of equal age. Dualism gets compromised because it cannot answer the question about the mediator signals that are necessary for the dual bodies to interact with one another. Suppose there are no signals; suppose the interaction is nonlocal. What then?

Human imagination and intuition reached such heights early on and formulated non-dualism or monistic idealism (also called *perennial philosophy*). God interacts with the world because God is not separate from the world. God is at once both transcendent and immanent in the world.

For the mind-body dualism, we can think idealistically in this way. Our internal experience, the abode of the mind, consists of a subject (that which experiences) and internal mental objects, such as thoughts. The subject experiences not only the internal objects, but also the external objects of the material world. Suppose we posit that there is only one entity, call it consciousness, which becomes split in some mysterious way into the subject and the objects in our experience. Consciousness transcends both matter and mind objects and is also immanent in them. In this way, the religious and philosophical languages become identical except for minor linguistic quibbles.

This philosophy of monistic idealism was never popular simply because transcendence is difficult to understand without the concept of nonlocality, a quantum concept. Even more obscure are such subtleties of the philosophy as stated in the sentence, "Everything is in God, but God is not in everything." The meaning of the sentence is that God can never be fully immanent; there is always a transcendent aspect of God. The infinite can never be fully represented in finitude. But try to explain that to the average person!

Nevertheless, monistic idealism has been very influential in the East, in India, Tibet, China, and Japan, in the form of religions such as Hinduism, Buddhism, and Taoism. These religions, not being organized hierarchies, always responded to the messages of mystics who from time to time reaffirmed the validity of the philosophy based on their own transcendent experience.

Mystics also existed in the West. Jesus himself was a great mystic. Following his lead, Christianity in the West has had other great mystics who have propounded monistic idealism, mystics such as Meister Eckhart, Saint Francis of Assisi, Saint Teresa of Avila, Saint Catherine of Genoa, etc. But the organized nature of Christianity drowned out the voices of the mystics (ironically, including Jesus), and dualism has prevailed in the official thinking of Christendom.

How do you recognize a mystic? These people have taken a quantum leap from their ego-mind to discover directly that there is existence, awareness, and bliss beyond the ego that is far greater in potential than what we ordinarily experience. But alas! The mystical breakthrough to a "more real" reality does not produce any immediate behavioral transformation (especially in the domain of base emotions). Therefore, behaviorally speaking, most mystics are usually no more impressive than ordinary people. We have to take the mystics' word for their "truth"—and scientists and social leaders through the ages have been reluctant to do that!

There is also a serious drawback to traditional philosophical formulations of monistic idealism. Everything is God or consciousness, so how real is matter, how important? Here most idealist philosophers take the view that the material world is irrelevant, illusory, only to be endured and transcended. True, a few idealist philosophers have emphasized the importance of the material by stating that only in the material form can one exhaust karma, which the soul must do in order to be delivered from the necessity of reincarnating time after time in physical form in the material world. But overall, there has always been an asymmetry in the outlook of idealists regarding consciousness and matter. Consciousness is the true reality, and matter is an epiphenomenon bordering on trivial. This is very similar to a reverse of the materialist belief that consciousness, mind, and all that internal stuff of our experience are trivial, lacking causal efficacy (a relation between one or more of the properties of a thing and an effect of that thing). For a complete, integral study of consciousness, we must rise above both of these attitudes.

EXTERNAL AND INTERNAL DOMAINS OF CONSCIOUSNESS, STRONG AND WEAK OBJECTIVITY

Obviously, the materialist studies of consciousness—neurophysiology, cognitive science, and so on—are limited by the belief system of the researchers, but no one can doubt that the data these researchers collect are useful. And the materialist theories, albeit incomplete, are useful too. Similarly, the data and theories garnered by mystics and meditation researchers through introspection of the internal, which leads to many reported higher states of consciousness (in addition to ordinary states), must also be regarded as meaningful and useful.

Recognize that what the materialist science studies is the third-person aspect of consciousness (behavioral effects), on which reaching a consensus is easy. The data satisfies the stringent criterion of strong objectivity—it is largely independent of the observer. In contrast, the mystics and meditation researchers study the first-person aspect of consciousness (felt experiences). We must realize that the data these latter researchers provide have similarities, and therefore they lead to a consensus about the higher states of consciousness. But we do have to relax the criterion for judging the data, from strong objectivity (observer independence: no subjective data are acceptable) to weak objectivity (observer invariance: the data are similar from one observer/subject to another). Note that typically in laboratory experiments of cognitive psychology, we already accept weak objectivity as the criterion for data on ordinary states of consciousness. Note also that, as the physicist Bernard D'Espagnat (1983) noted long ago, the probabilistic nature of quantum physics is consistent only with weak objectivity.

We can add to this summation another quadrant, the intersubjective experience—the scantly studied data on internally experienced aspects of relationships. And to make it all symmetric, we can add a fourth quadrant, consisting of objective data about conglomerates of people, such as entire communities. In this way, we get the four-quadrant model (figure 3-1), thanks largely to the philosopher Ken Wilber (2000).

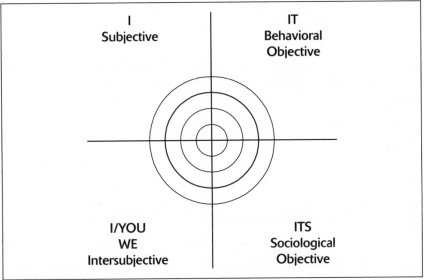

I Subjective	IT Behavioral Objective
I/YOU WE Intersubjective	ITS Sociological Objective

FIGURE 3-1. The four quadrants of consciousness according to Wilber.

However, while this phenomenological coup may seem like an integrated approach, in truth it is just a beginning. Dichotomies remain in each quadrant; also, no real integration of all the quadrants has been achieved. The philosopher's position is elitist: one cannot integrate using reason or science. To see the integration, one has to achieve higher states of consciousness.

Can we overcome the philosopher's prejudice that science applies only to the material level of reality and reason can never be extended to treat higher levels of consciousness? I think that this prejudice originated in the philosopher's belief in a hidden dualism of consciousness and matter, of interior reality and exterior reality. The philosopher then tries to avoid the problem of interactionism (how consciousness and matter interact) by claiming that science applies only to the exterior (matter) and not to the interior (consciousness), so that there is no need to bother about how the two interact.

When the true meaning of quantum physics is understood, it becomes clear that consciousness cannot be a mere phenomenon of the brain. Furthermore, there is no need to undermine mind and other internal objects as epiphenomena of the brain and the body. Instead, quantum physics and all science must be based on the philosophy of monistic idealism: consciousness is the ground of all being, in which matter, mind, and other internal objects exist as possibilities. But there is no reason to undermine matter either. Matter in its capacity to represent subtle mental states is as important as the subtle (nonmaterial) that it reflects. In other words, quantum thinking allows us to treat mind and matter, internal and external experiences on equal footing, extending causal efficacy and importance to both.

In this way, philosophically and scientifically (with theory and evidence), we have solved the metaphysical problem of which "ism" is accurate and valid—monistic idealism. However, materialist thinking has created a wound in the collective psyche of humanity that, unattended and unhealed, is only getting worse. Our primary job now is to help heal this wound by sharing the philosophical and scientific message of unity that is emerging with all of humanity.

As modernists, we have acknowledged the veracity of the mind and what it processes: meaning. This has led to a much more expansive participation in the adventurous exploration of meaning. As modernism has given way to the postmodern malaise of meaningless materialism, our institutions and their progressive legacy of democracy, capitalism, and liberal education have been put in jeopardy. They are being undermined to create a new kind of hierarchy, setting new limits on freedom that are no better than the limits imposed in the past by church and feudal domination. This time, the shackles are materialist science and scientism.

Monistic idealism can lead to a new kind of modernism that I call *transmodernism*, following the philosopher Willis Harman. Descartes' dualistic modernism was based on the motto, "I think, therefore I am." In other words, if there is a thought, there must be a thinker. This released the thinking mind for new exploration, but mainly for inventions intended to solve problems. Inventions require creativity, but only a limited version of it that I call *situational creativity*, which is designed

46

to solve a problem within a known context of thinking. Situational creativity is important, but in some real sense, it is also more of the same: it is "thinking inside the box." Transmodernism is based on the motto, "I choose, therefore I am." It releases the true potency of the creative mind, not only situational creativity but also what I call *fundamental creativity*: the ability to change the contexts on which thinking is based and choose new ones.

Under modernism, we got not only the benefits of democracy and capitalism, but also the evils of modernism: thinking that put humans over nature and the domination of thinking over feeling, which I call the *mentalization of feeling*. Yes, we have created useful industry and technology, but we have also created environmental problems that we don't know how to solve.

We need to bring back the modernist spirit and the emphasis on mental exploration, but without its dark side, its attitudes of human-over-nature and reason-over-feeling, and without the almost total dependence on simple hierarchies and the ego isolation of the lone individual. The new era of transmodernism begins with a quantum leap in our attitudes—from human over nature to human within nature, from reason over feeling to reason integrated with feeling, from simple hierarchies to tangled hierarchies, from ego separateness to the integration of the ego and quantum consciousness/God. Then we are truly back on track for the emergence of a new age of ethical living.

OLD SCIENCE AND THE NEW SCIENCES: PARADIGM SHIFT

I introduced the idea of a paradigm shift in science in chapter 1. The old science is based on the supremacy of matter, material monism, with its reductionism and upward causation. The new holistic paradigm does not give up the material monism: everything is matter. But it does give up the idea of reductionism and opts for the philosophy of holism: the whole is greater than its parts and cannot be reduced to its parts. Here God and spirituality are recovered in the sense of an immanent God, or a "Gaia consciousness" immanent throughout the whole world with all its organisms. (The Gaia hypothesis or theory, developed by James

47

Lovelock, represents all things on earth, living and nonliving, as a complex system of interactions that can be considered to be a single organism.) There is also something like downward causation, a causal autonomy of the emergent holistic entities at each level of organization that cannot be reduced to the parts. Alas! This causality is not real, because in the final reckoning it too is determined from material interactions, that is, from upward causation.

The newest science, science within consciousness, is based on quantum physics and the primacy of consciousness (monistic idealism), and it is inclusive of the old reductive paradigm. In science within consciousness, God is a real, causally efficacious agency, intervening through downward causation. In science within consciousness, we can even treat subtle bodies without the usual problems of interaction dualism. In science within consciousness, we can address within science the evolution of godliness that religions aspire to achieve. And yet the old science remains valid—in its own domain. In the material domain of conscious experience, consciousness chooses the actual event of manifest reality out of the quantum possibilities determined by the upward causation from the material substratum. And since quantum effects are relatively muted for gross matter, gross material behavior is approximately deterministic.

In truth, even reductionistic materialists make some room for God. In a book called *Why God Won't Go Away,* Andrew Newberg and Eugene D'Aquili (2001) cited recent work in neurophysiology to suggest that God and spiritual experiences can be explained simply as brain phenomena.

In a similar vein, the holists maintain that God and spirituality can be understood and explored as an emergent holist phenomena of matter itself; even free will and downward causation can be understood as emergent apparent autonomy of higher levels of organization of matter.

The paradigm explored and endorsed in this book is much more radical than either of these two approaches to God. I posit that the ground of being is consciousness, not matter. I posit that not only matter but also a subtler vital energy body, an even subtler mind, and an even subtler supramental body all exist as quantum possibilities of consciousness. These develop in time from causal interactions in their respective domains. I also posit that as we evolve we move through manifest states of consciousness that are greater and greater manifestations of godliness—the qualities of God, the supramental archetypes. The price we pay for including the subtle in our science is multiculturalism of theory and weak objectivity for sorting data.

I must emphasize once again that the God for which I present scientific data is quite the same as the God envisioned by mystics and the founders of all of the world's great traditions, although the teachings of the great religions have become diluted into dualism in their popular renditions.

Toward the end of the 19th century, the philosopher Nietzsche pronounced through one of his fictional characters that "God is dead." This reflected Nietzsche's uneasiness about the effectiveness of a naïve popular Christianity to uphold ethics and morality against the materialist science that was rapidly sweeping the West. In other words, Nietzsche realized that the popular dualistic Christian God was dead. I show in this book that in the new paradigm of science based on the primacy of consciousness and quantum physics, God lives on eternally as the agent of downward causation, in a role that should prove satisfactory to both science and religion.

Irrespective of whatever picture of God currently satisfies you the most, I hope you will give the evidence and theory presented here a fair appraisal. After all, God has been a preoccupation of human beings for millennia, a preoccupation that I suspect has affected you at least a little. I merely ask that you suspend your judgment and disbelief while you read Parts Two, Three, and Four, in which I present the evidence.

Chapter 4

God and the World

Our old science tells us that what is real is the material universe, that our individual brains are material, and that our experience of our egos and of God are merely epiphenomenal experiences of these brains. Some mystics say that only God is real, and that the manifest world is unreal. The popular version of Christianity tells us that both the material universe and God are real, but they are separate realities.

The new science tells us that the universe, God, and we are not really separate: the separateness of God, the world, and us is an appearance, an epiphenomenon.

So what is real and what is epiphenomenon? That is the question. Or is it?

If our consciousness is unreal, there is no reason for me to write this book or for you to read it. So why do we—materialists included—read, write, research, and want to know reality, and even love and want to be happy? Because in our hearts, we *know* that our consciousness *is* real, that it has causal efficacy. There is vitality in our feelings, there is meaning in our conscious thoughts, *and there are purpose and value in our intuitions.* As Descartes argued long ago (using a slightly different language), we can question the reality of everything else, but not of our consciousness. In the same vein, is the material unreal? If you talked about the unreality of the material world to Zen masters of old times, those connoisseurs of subtler states of consciousness, they might have pulled your ear and, when you would complain, asked, "Why are you complaining about what is unreal?"

The real questions are these: Why does the world appear to be separate from us? What does the fact that we get lost in this separateness from the universe and from each other do to us and to the human condition? Is there any way to go beyond this dynamic of separation?

In the new science, we find that the world is here because of us and that we are here because of the world. (See chapter 7.) The separation dynamic is one of mutual creation, our prerequisite for play in manifestation. When we creatively comprehend this, the separation dynamic loses its hold on us. The story of the universe is our story. When we understand ourselves, our consciousness, we also understand our relationship with the universe and with God, and the separation becomes a portent for play.

What happens to this sense of manifest play when the separateness is seen as illusory? I hope you are curious to find out. I hope you are tired of the old play of real separateness, which has given us the nightmares of terrorism, energy crisis, global warming, and the possibility of nuclear war. I hope you are ready to explore the potential of a science within consciousness, as well as the potential of waking up to subtler levels of consciousness. I hope you are ready to appreciate the importance of the scientific rediscovery of God.

Many of the present inquiries of materialist science sound more like the medieval question of how many angels can dance on the head of a pin. Does knowing every little detail about a black hole give you any inkling about how to love or forgive? The old tired science cannot give us answers to our big problems. Why is there so much terrorism and how do we deal with it? Why is there so much violence? And how do we deal with our children committing mass murders? Why is there so little love? How do we reintroduce ethics into our society and love into our families? Are ethics important? If so, how do we teach ethics and values to our children when our pervasive materialist science professes that the world is value-free? Why the ongoing economical ups and downs with capitalism? How do we produce steady-state economies? How do we transform capitalist economics so that the gap between the rich and the poor becomes narrower instead of wider, so that even the poor can use their minds to process meaning? How do we make our business and industry ecology-friendly, so that we can protect our planet against global warming and other ecological disasters? Why are politics so corrupted? How do we defend democracy against the power of money, the media, and the fundamentalists? How do we curb the skyrocketing cost of health and healing?

Like the question of consciousness, these are hard, even impossible questions for the materialist worldview. But materialists go on claiming that the answers are right around the corner, an attitude that the philosopher Karl Popper called *promissory materialism*. It is only human to stick our heads in the sand when hard questions are asked; materialist scientists are no exception. But while these scientists deplore the "George Bush syndrome" in the case of global warming, they turn around and display the same attitude when it comes to acknowledging that a paradigm shift is necessary and inevitable in order to include consciousness in our science and worldview.

Meanwhile, just as global warming is endangering our world, urgent social problems are growing that cannot be solved within the materialist approach; in fact, for most of these problems, materialism is the root cause. And of course, there are also those age-old subjective human questions that materialist science avoids: What is the meaning of my life? How can I love? How do I become happy? What is the future of my evolution? These questions are impossible for the materialist, but the new science within consciousness allows us to make a good beginning in finding answers to them.

But can we put God/consciousness back into our systems of knowledge to change our behavior and societies in time to avoid the catastrophes that are threatening us? Yes, we can.

I will tell you my hypothesis: even those influences that have led us today to near cataclysm are part of a purposive movement of consciousness that is already under way to avert these catastrophes.

Meanwhile, the looming catastrophes are trying to tell us something important; this we have to decipher. We have to recognize the meaning and purpose of the movements of consciousness. Then the choice is ours. Do we align ourselves with the evolutionary purpose of those movements, do we run against it, or do we play apathetic?

You also have to recognize the one common aspect of all the catastrophic problems—conflict. Current terrorism has its root in the conflict between materialism and religion. It is not only the fundamentalist Muslims of the Middle East who are fighting the materialistic "Great Satan" empire of the United States, but also Christian fundamentalists within this country. Economic and ecological problems are superficially due to the conflict between individual and collective interests, to ego values (such as selfishness and excessive competition) and being values (such as cooperation, win-win philosophy, intuition, creativity, feeling, happiness). Ultimately these causes too can be traced to the conflict between materialism and spirituality. On close examination, the major reason that health costs are rising is our fear of death and the conflict between ignorance and wisdom—again, materiality and spirituality. The decline of ethics and values in our families, our societies, and our schools is clearly due to this conflict. To enter the world of true solutions is to resolve the conflict.

INDIVIDUAL AND COLLECTIVE MOVEMENTS OF CONSCIOUSNESS

Spiritual traditions of the East understand the individual movements of consciousness very well, and that is what they emphasize. Living in the world produces an individual identity (ego) superimposed upon the cosmic God-consciousness. This is the ignorance that obscures the wisdom of oneness. Easterners believe that through many incarnations, the ego identity gives way to God-consciousness, and knowing that one is God, one is liberated from the cycle of birth, death, and rebirth. Hence the adage from this point of view: *you cannot change the world, you can only change yourself.* Whatever change the world undergoes will come through these individual changes.

But in the West, the belief in only one life has undermined the drive for self-realization and transformation. Instead, the emphasis has been on ethics: following certain rules of behavior to bring oneself in alignment with God. Even under the aegis of materialism, the West has developed a social consciousness in which there clearly is some imperative for social ethics:

There is only one life and it is too short
Let's work together and improve our lot.

So today we also have activists who try to change the world, but often have no spiritual notion of changing themselves. Can we see the necessity of both trends and integrate them?

EVOLUTION

In both the East and the West, whether we believe in reincarnation or in one life, the emphasis of spirituality has been to unite with a transcendent God. Spiritual philosophers, of course, are quite aware that God is also immanent in the world, but somehow they more or less have managed to undermine our pursuit of that unity in the immanent world. To some extent, this has contributed to making the world culture materialistic. More recently, spiritual traditions have allowed the affairs of the world to be dominated by materialist science, which until recently has propagated materialism in the world without any challenge. Only in the last few decades has a challenge to materialism surfaced from within the tradition of science itself.

In dualist cultures, spiritual philosophers have wondered why a perfect God creates an imperfect world. In nondualist cultures, spiritual philosophers have occasionally wondered aloud why God is immanent in an imperfect world when He could have forever stayed in heavenly perfection. The answer to both concerns is, of course, evolution. In both cultures, spiritual thinkers have missed evolution. God becomes manifest in the immanent world to manifest its unmanifested possibilities. The world begins the journey of evolution imperfect, but that is only a beginning. Consciousness evolves toward perfection in manifestation, toward seeing its perfect nature in manifestation.

55

Because spiritual traditions have neglected worldly affairs in general, it is not surprising that they have not recognized evolution to be a major part of the manifest play of consciousness. Coincidentally, materialist science, which discovered evolution (and used it to obstruct the influence of religion in society), has not seen evolution as a major force in our life either. More or less, biologists are content with an inadequate theory of slow and gradual Darwinian evolution. In Darwinism, evolution is posited to occur in two steps. First, variations occur in the hereditary components (the genes) of a species; second, members of the species that survive and reproduce in greater numbers will pass along their genes through what is known as *natural selection*, giving the species a better chance of surviving. In this way, although evolution is seen as relevant to our survival, it has no other importance. If it would improve the chance of survival for the human species to become less complex, less oriented toward meaning and values, that direction of evolution would be OK with the Darwinists. In short, evolution is all about physical survival, not spiritual development.

But this, too, has been changing. The empirical persistence of discontinuities in the geological record, *fossil gaps,* has made it clear that (neo-) Darwinism, which predicts continuous evolution, is an inadequate theory and that we must invoke downward causation and biological creativity for a complete theory of evolution (Goswami, 1997a, 2008). In the new approach, evolution is recognized as purposive and as a major force in our lives.

In the last century, two philosopher-sages, Sri Aurobindo in India and Pierre Teilhard de Chardin in the West, had the revolutionary insight that evolution does not end its journey toward increasing complexity with humans. According to Aurobindo, just as animals have been the laboratory for nature to evolve humans, similarly human beings are currently the laboratory for evolving superhumans. And in superhumans, we will see the heavenly qualities that we strive to attain—love, beauty, justice, good, etc.—come forth and evolve toward perfection. The end of evolution is when we reach the omega point of perfection, according to Teilhard.

We have to recognize that evolution is also a play of consciousness—a purposive, collective play. The collective movement toward social consciousness that started in the West (mainly) is *important and an essential and integral part of the movement of manifest consciousness*— the evolution of consciousness. So here again the Eastern and Western views of life must be integrated. We have to proceed toward individual salvation as in the East, but we also *have to contribute to evolution*. And clearly, we add more to the movement of evolution as we shed our ego identity in our journey toward God.

We need a new kind of activism with a new adage: you cannot change the world, but you can change yourself, always with the perspective of *collective world evolution* in mind. This is what I call *quantum activism,* in which we work on transforming ourselves using the power of the new physics, but while paying attention to the evolutionary movement of consciousness as a whole, always trying to heed its needs.

Religions have traditionally encouraged us to follow God for personal salvation, for alleviation of our suffering, for the discovery of effortless happiness in living. But for most types of suffering, we now have antidotes (albeit temporary ones). It is no longer clear why we should make arduous efforts to establish God in our lives, to embrace additional suffering for some elusive happiness in the future. Some people among us, of course, still do it anyway, and we wonder why. What motivates them? I submit that the motivation for finding God in our lives comes from an evolutionary pressure for those ready to move beyond their ego-boundary. The very existence of this pressure suggests that we are preparing for a new evolutionary stage. This is what was foreseen by Aurobindo and Teilhard.

So What Is Our Response to Evolution?

The time has come, I declare, to acknowledge the rediscovery of God within science. If it requires a paradigm shift of our science from its base of matter to a base of consciousness, so be it. We must also proceed to actualize the God potency within us, as far as each of us is capable, if we are interested in the welfare of the world.

I suggest that you can do a lot of things to begin a journey from separateness to unity, from ego-consciousness to God-consciousness, and from stasis to evolution. Here are some starters:

> Think quantum! Think possibility!
>
> Explore the potential of consciousness.
>
> Explore the possibility that your separateness from the whole is illusory; study the nature of your conditioning.
>
> Practice and realize the power of freedom of choice.
>
> Pay attention to your inner experiences and your subtle bodies in addition to the outer and the gross.
>
> Resolve conflicts and integrate partial (and conflicting) bits of wisdom into a whole.
>
> Prepare to wake up to the nonlocality of consciousness.
>
> Recognize the importance of working on your own transformation while acknowledging the evolution of consciousness (movement of the whole). Pay attention to the movements of consciousness as they pertain to our social institutions.
>
> Move from the world of impossible problems (materialist science) to the world of possible solutions (science within consciousness).

THE PLAN OF THE BOOK

I hope that the Part One of the book has given you a good introduction to the God that we are rediscovering in science and shown you in what way the scientific God is different from the God of pop religions. But please note that in the most fundamental basics—downward causation, the existence of subtle bodies, and the importance of godliness—there is agreement. This agreement is most important, and I hope it will encourage further dialogue between the (new) science and dualistic popular religions.

As promised in chapter 1, Parts Two, Three, and Four present the new evidence in support of the existence of God.

In Part Two, I introduce the nature of the quantum signatures of the divine in some detail and expound on the experimental verification of downward causation, quantum nonlocality, discontinuity, and tangled hierarchy in psychology and in biology. This includes an explanation of the distinction between the conscious and the unconscious and between life and nonlife. Part Two ends with a discussion of creative evolution, a God-based theory of evolution that explains the fossil gaps and the why and how of intelligent design.

Part Three consists of the theory and experimental proof of the existence of the subtle bodies; these aspects of the subject allow us a very timely extension of science to tackle the "impossible" problems of biological and psychological sciences. These include questions about the nature and origin of our feelings of being alive, about the validity of homeopathy and acupuncture, and about the value of divining phenomena such as dowsing.

Part Three is also about the new psychology and how it explores the "mind of God." I explain why we have both inner and outer experiences and why both are important for science to validate and expound on the notion that "my father's house has many mansions" (*Bible*, John 14:2). Is God up above, down here in immanence alone, or down under? Questions like these shed new light on how to investigate and know God. Data on dreams, states of consciousness, reincarnation, ethics, and altruism tell us about the soul.

Creativity, love, transformation, and healing are examined as examples of divine downward causation in Part Four. They are all shown to provide irrefutable evidence for the existence of a transcendent God.

Part Five deals with quantum activism and what we can do to evolve ourselves and our society in accordance with the evolutionary agenda of consciousness. Here I also discuss how to unleash the transformative power of quantum physics in our journey of quantum activism.

The book ends with two special epilogues. The first addresses the young scientist confused by the claims of materialist science that do not add up. The second shows that Jesus, the father of Christianity, was quite tuned to the lessons of quantum physics. He knew.

IN SUMMARY

There are aspects of the phenomenal world that are impossible to address within the materialist view of science (only one level of reality— material—and only one source of causation—upward):

It is impossible to collapse quantum possibility waves to actual events.

It is impossible to explain quantum measurement—the collapse of the quantum possibility wave into an actual event.

It is impossible to explain discontinuity in terms of only continuous operations.

It is impossible to generate nonlocality from local interactions alone.

It is impossible to bring about circular, tangled hierarchy from linear simple hierarchies.

It is impossible to distinguish between conscious awareness (the subject-object split of an experience) and the unconscious (no subject-object split awareness).

It is impossible to distinguish life from nonlife.

It is impossible to explain the interior experience (the first-person subjective) in terms of the exterior (the third-person objective).

It is impossible to explain the processing of meaning in terms of symbol processing capacity.

It is impossible to explain feeling from symbol processing capacity alone.

It is impossible to explain the laws of physics from material movement alone.

The incompleteness and inadequacy of our dominant paradigm of science show up clearly when we encounter phenomenon after phenomenon that is impossible in the materialistic scheme. These constitute real, impossible gaps in our understanding that a materialist science can never bridge, even in principle. It is in these impossible-to-explain gaps that God is rediscovered.

To paraphrase Shakespeare, there are more things in heaven and earth, oh materialist, than are dreamt of in your philosophy. Acknowledge it!

This reminds me of a story about Mulla Nasruddin, a figure from the 13th century, the subject of stories throughout the Middle Eastern world. He was found working vigorously with a pail of water, shaking the water, beating it with his hands, kneading it, creating quite a spectacle.

Somebody asked, "Mulla, what are you doing?"

To this, Mulla said, "I am making yogurt."

The questioner was shocked. "Mulla, you can't make yogurt out of water!"

Mulla replied, "And what if it works?"

The difference between the usual gap theology (see glossary) and the scientific approach presented here is, of course, that we are not content with merely suggesting God as an explanation for the gaps in materialist science. Instead, we build a new verifiable science based on the God hypothesis with experimental evidence.

How can God be said to have been rediscovered in science? Because we have a verifiable scientific theory based on the God hypothesis that explains in full scientific detail how the impossible becomes possible, how the gaps are bridged. And most important, some of the crucial predictions of this theory have already been verified in scientific experiments. In the years to come, we can look forward to further laboratory verification of this new science.

So, have I succeeded in providing enough scientific evidence for the existence of God? For some people, especially the religious fundamentalists, unless you are providing evidence for their God that satisfies their theology, no evidence will be enough. Similarly, the hardcore materialist will not be persuaded by any amount of new data, new verifiable predictions, new explanations of old puzzles, or new resolutions of impossible paradoxes. But in between these two extremes, there are many people, laypersons and professionals alike, scientists and nonscientists, who will appreciate what I have presented—for the simple reason that never before has it been possible to integrate so many disparate scientific fields and notions with so very few new assumptions, the primary one being God as quantum consciousness.

If you have read this far in this book, you are one of those people. And it is up to you to judge if this book helps you in your journey to God or evolution or both. I have done my best to provide you with concepts to research and to internalize, maps for the journey to view, questions to ponder, intentions to arrive at, and jobs to do.

Part Two

The Evidence for Downward Causation

I n 1979, I had found my "happy" physics: the "quantum measurement" problem, how quantum possibilities become actual events in an observer's experience in the process of merely looking. Whenever I thought about the problem and its possible solution, I would be puzzled. But strangely, it made me happy. I was positive I was doing something that would result in "disturbing the universe."

The famous physicist John von Neumann had left us with a hint: it is the observer's consciousness that changed the waves of possibility of a quantum object into actual events, the particles that one saw. But what is consciousness?

Nobody knew. The conventional ways of thinking about it gave paradoxes when applied to the measurement problem. A graduate student in physics suggested I seek out psychologists, since they study consciousness. So for a few years I collaborated with a psych professor and learned the psychological perspective of consciousness.

No answers emerged that resolved my paradoxes. The physicist David Bohm was becoming prominent at the time. I started reading him and noticed that Bohm was hobnobbing with the mystic J. Krishnamurti. What did I have to lose? I started hobnobbing with mystics.

In May 1985 I was visiting a friend in Ventura, California, and we attended a Krishnamurti talk in nearby Ojai. After the talk, we were settled down in my friend's living room with a mystic named Joel Morwood.

Soon the conversation turned to New Age science. I explained to Joel how paradoxical it was that consciousness, no doubt an emergent phenomenon of the brain, nevertheless could "collapse" the quantum possibility waves of all the objects we see, including those in the brain.

And Joel challenged, "Is consciousness prior to the brain or is the brain prior to consciousness?"

I knew that mystics put consciousness prior to everything. So I carefully said, "I am talking of consciousness as the subject of experiences."

"Consciousness is prior to experience," said Joel. "It is without an object and without a subject."

I knew those phrases, too. Just a while ago, I had read a book by the mystic-philosopher Franklin Merrell-Wolff (1983) entitled The Philosophy of Consciousness Without an Object.

So I countered, "Sure, that is vintage mysticism, but in my view you are talking about the nonlocal aspect of consciousness."

It was then that Joel gave me an emotional little lecture about how I wore "scientific blinders." Those were his exact words. He ended with the Sufi statement, "There is nothing but God."

Now mind you, I had heard or read those words many times by then, in different contexts from different traditions, but this umpteenth time, understanding dawned and a veil lifted. I suddenly realized that the mystics are right—consciousness is the ground of all being, including matter and brain, and that science must be built on this metaphysics rather than on the traditional materialist metaphysics.

It took me four years to publish my first paper (Goswami, 1989) with a paradox-free solution of quantum measurement problem. In that paper I was careful to mention neither mystics nor any mystical literature, let alone "God." I was scared that if I did, scientists would reject my thesis outright.

Four years later, when I was writing my book The Self-Aware Universe, *I was not as self-conscious. And when Jacobo Grinberg Zylberbaum invited me to the University of Mexico in 1993 to look at his experimental setup and the data, while we were writing the paper on his experiment, I knew. I knew that we were rediscovering God in science.*

Chapter 5

The Quantum Signatures of the Divine

Jesus lamented that the kingdom of God is everywhere, but people don't see it. Well, the evidence is subtle; it is easy for ordinary people to miss it. But scientists are special people; they are experts in deciphering subtleties of evidence. Why have they been missing the signatures of the divine?

The Nobel laureate physicist Richard Feynman expressed this myopia of the scientists of recent times when he offered this admonishment against unbridled imagination. He said, "Scientific imagination is imagination within a straightjacket." The straitjacket Feynman and other materialists wear is the belief system called *scientific materialism*, which I have already mentioned. And the doctrine that binds the most is the exclusivity of the reductionist's doctrine of upward causation.

This entire book is an exercise in how to get free from the straitjacket of materialism. In chapter 1, I argued that quantum physics is showing us the way by giving us back downward causation and its agent: God acting through the observer. In Newtonian physics, objects are determined things. But in quantum physics, objects are possibilities from which consciousness chooses. When a person looks, his or her consciousness chooses among the quantum possibilities to collapse an actuality of experience.

But how is this evidence for the existence of God? It sounds like a Pogo cartoon: we have searched and found God—and it is us! Maybe the ancient Hindus were right when they said there are 330 million Gods. Well, it is six billion now because of inflation. If we are God, why do we live the way we do? Why do we have such a hard time manifesting godly qualities like nonviolence and love?

The evidence for God is within us, but to see it we have to be subtle. To live it, we have to grow.

WE CREATE OUR OWN REALITY, BUT ...

It was in the 1970s that the physicist Fred Alan Wolf (1970) created the evocative phrase "we create our own reality." The images the phrase evoked led, however, to many disappointments. Some people tried to manifest Cadillacs, others vegetable gardens in desert environments, and still others parking spaces for their cars in busy downtown areas. Everybody was inspired by the idea of the quantum creation of reality, no doubt, but the attempts at creation produced a mixed bag of results because the would-be creators were unaware of a subtlety.

We create our own reality, but there is a subtlety. We do not create reality in our ordinary state of consciousness, but in a non-ordinary state of consciousness. This becomes clear when you ponder the paradox of Wigner's friend, a thought experiment proposed by the Nobel laureate physicist Eugene Wigner, who first thought of the paradox. Here I present the paradox with a simple example.

Imagine that Wigner is approaching a quantum traffic light with two possibilities, red and green, and that at the same time his friend is approaching the same light on a cross street. Being busy Americans, they both choose green. Unfortunately, their choices are contradictory: if both choices materialize at the same time, there would be pandemonium. Obviously, only one of their choices counts—but whose?

After many decades, three physicists in different places at different times—Ludvik Bass (1971) in Australia, I (Goswami, 1989, 1993) in

Oregon, and Casey Blood (1993, 2001) in New Jersey—independently discovered the solution to the paradox: consciousness is one, nonlocal, and cosmic, behind the two local individualities of Wigner and his friend. They both choose, but only figuratively speaking: the one unified consciousness chooses for both of them, avoiding any contradiction. This allows the result dictated by quantum probability calculations that, if Wigner and his friend arrived at the same traffic light on many occasions, each would get green 50% of the time, yet for any one occasion, a creative opportunity for getting green is left open for each.

In 2003, I was invited to give a talk at a scientific conference on consciousness in London. After my talk, a BBC reporter had a question for me: "Does your theory prove the existence of God?" I saw the trap in his question immediately. If I said yes, he would have a sensational headline for his report, "Quantum physicist supports the idea of God sitting on a majestic throne in heaven doling out acts of downward causation." So I said cautiously, "No and yes." He seemed a little disappointed that I did not fall into his trap. I elaborated. No, because the God rediscovered by quantum physics is not the simplistic God of popular religions. God is not an emperor in heaven doling out downward causation, judgments as to who is to go to heaven and who is bound for hell. Yes, because the author of quantum creation, the free agent of downward causation, transcends our ordinary ego. It is universal and cosmic, exactly like the creator God posited by all the esoteric traditions of spirituality. You can call It quantum consciousness, but Its flavor is uniquely that of what the traditions call God.

The oneness of the choosing consciousness is an outcome of the question we pose: What is the nature of consciousness that enables it to be the free agent of downward causation without any paradox? For one thing, consciousness has to be unitive, one and only for all of us. This oneness of consciousness is then a prediction of the theory.

When my paper (Goswami, 1989) was published proclaiming this prediction in an obscure physics journal, University of Mexico neurophysiologist Jacobo Grinberg-Zylberbaum noticed it. Jacobo had been doing experiments with pairs of human subjects and strange transfers of electrical brain activity between them. He intuited that my theory may have something important to add to the interpretation of his experiments. So I got an excited call from him. To make a long story short, I flew out to his laboratory at the University of Mexico, checked out his experimental setup and the data, and helped him interpret it. And in a short while, Grinberg-Zylberbaum and three collaborators (1994) wrote the first paper proclaiming a modern scientific verification of the idea of oneness of consciousness.

THE GOOD NEWS EXPERIMENT: WE *ARE* ONE

The good news is that four separate experiments are now showing that quantum consciousness, the author of downward causation, nonlocal, unitive, is God.

As mentioned above, the first such experiment proving it unequivocally (with objective machines and not through subjective experiences of people) was performed by the neurophysiologist Grinberg-Zylberbaum and his collaborators at the University of Mexico. Let's go into some details.

Quantum physics gives us an amazing principle—nonlocality. The principle of *locality* says that all communication must proceed through local signals that have a speed limit. Einstein established this limit as the speed of light (the enormous but finite speed of 299,792,458 m/s). So this locality principle, a limitation imposed by Einsteinian thinking, precludes instantaneous communication via signals. And yet, quantum objects are able to influence one another instantly, once they interact and become correlated through quantum *nonlocality*. This was demonstrated by the physicist Alain Aspect and his collaborators (1982) for a pair of photons (quanta of light). The data are not seen as a contradiction to Einsteinian thinking, once we recognize quantum nonlocality as signal-less interconnectedness outside local space and time.

Grinberg-Zylberbaum, in 1993, was trying to demonstrate quantum nonlocality for two correlated brains. Two people meditate together with the intention of direct (signal-less, nonlocal) communication. After 20 minutes, they are separated (while continuing to meditate upon their intention) and placed in individual Faraday cages (electromagnetically impervious chambers), where each brain is wired up to an electroencephalogram (EEG) machine. One subject is shown a series of light flashes producing electrical brain activity that is recorded by the EEG machine. From this record, an "evoked potential" is extracted with the help of a computer (upon subtraction of the brain noise). This evoked potential is somehow transferred to the second subject's brain, as indicated by the EEG record of this subject, which gives (upon subtraction of noise) a potential similar in phase and strength to the potential evoked in the first subject. This is shown in figure 5-1. Control subjects (who do not meditate together or are unable meditatively to hold the intention for signal-less communication during the experiment) do not show any transferred potential (figure 5-2).

The experiment demonstrates the nonlocality of brain responses to be sure, but also something even more important—nonlocality of quantum consciousness. How else can we explain how the forced choice of the evoked response in one subject's brain can lead to the free choice of an (almost) identical response in the correlated partner's brain? As stated above, the experiment has been replicated several times—first, by the neuropsychiatrist Peter Fenwick and collaborators (Sabell *et al.*, 2001) in London; second, by Jiri Wackermann *et al.* (2003); and third, by the Bastyr University researcher Leanna Standish and her collaborators (Standish *et al.*, 2004).

The conclusion based on these experiments is radical. Quantum consciousness, the precipitator of the downward causation of choice from quantum possibilities, is what esoteric spiritual traditions call God. We have rediscovered God within science. And more. These experiments usher a new paradigm of science based not on the primacy of matter, like the old science, but on the primacy of consciousness. Consciousness is the ground of all being, which we now can recognize as what the spiritual traditions call Godhead (Christianity), Brahman (Hinduism), Ain Sof (Judaism), Shunyata (Buddhism), and so on.

The new science integrates. Whereas most of these terms that denote the ground of being, Godhead for example, indicate its fullness, the Buddhist term Shunyata indicates a void or nothingness. Contradiction? The new science explains: the ground of being is full of

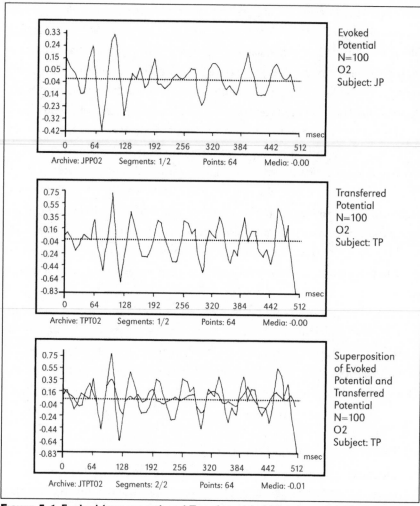

FIGURE 5-1. Evoked (uppermost) and Transferred (middle) potential. The bottom curve shows a 71 percent overlap between the two (from Grinberg-Zylberbaum et al., 1994).

possibilities, yes, but possibilities are not "things," so it can also be correctly called "no thing-ness."

THE POWER OF INTENTION

One of the most important aspects of the Grinberg-Zylberbaum experiment is demonstrating the power of our intention. His subjects

70

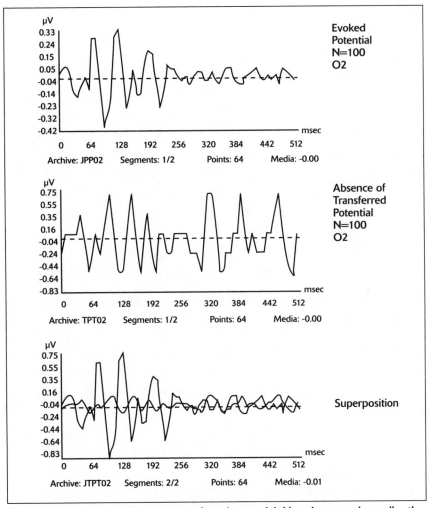

FIGURE 5-2. Control subjects: no transferred potential. Note how much smaller the observed potential for the second subject is (middle curve; pay special attention to the vertical scale). Also, the bottom curve shows no appreciable overlap.

intended that their nonlocal connections manifest. The parapsychologist Dean Radin (1997, 2006) has done more experiments demonstrating the power of intention.

One of his experiments took advantage of the O. J. Simpson trial in 1994-1995. At that time, lots of people were watching the televised trial and Radin correctly hypothesized that their intention watching the trial would fluctuate widely depending on whether the courtroom drama was intense or ho-hum. On the one hand, he had a group of psychologists make a plot of the intensity of the courtroom drama (and hence the intensity of people's intentions) as a function of real time. On the other hand, in the laboratory he measured the deviation from randomness of what are called *random number generators* (which translate random quantum events of radioactivity into random sequences of zeroes and ones). He found that the random number generators deviated from randomness maximally at precisely those times when the courtroom drama was the most intense. What does this mean? The philosopher Gregory Bateson has said, "The opposite of randomness is choice." So the correlation proves the creative power of intention.

In another series of experiments, Radin found that random number generators deviate from randomness in meditation halls when people meditate together (showing high intention), but not at a corporate board meeting!

The inquisitive reader is bound to ask about how to develop the power of intention. The fact is, we all try to manifest things through our intentions; sometimes they work, but more often they do not. Now we see that this is because we are in our ego when we intend. But how do we change that?

This is a very good question. An intention must start with the ego; that is where we ordinarily are—local, individual, and selfish. In the second stage, we intend for everyone to achieve what we want to achieve; this is to go beyond selfishness. We don't need to worry; we haven't lost anything—when we say "everyone," that includes us, too. In the third stage, we allow our intentions to become a prayer: "If my intention resonates with the intention of the whole, of God, then let it come to fruition." In the fourth stage, the prayer must pass into silence and become a meditation. This is important because only in silence can the possibilities to choose from grow.

If you seriously practice this, don't expect overnight results. Today, with our busy lifestyles, silence is difficult for us. Grow silence. Slow down your lifestyle. Make room for new possibilities. Then, manifest your intention, discontinuously. This is the real secret of manifestation.

DISCONTINUITY AND QUANTUM LEAP

Downward causation occurs in a nonordinary state of consciousness that we call *God-consciousness*. Yet we are unaware of it. Why are we unaware? Mystics have been telling us about the oneness of God-consciousness and our ordinary consciousness for millennia, but we haven't heard them for the most part. Why?

The *Upanishads* of the Hindus say emphatically, "You are That," meaning you are God! Jesus said, no less emphatically, "You are all the children of God." This is a key. We are children of God; we have to grow up to realize our God-consciousness. There are mechanisms (see below) *that obscure our Godness*, giving rise to our ordinary I-separateness that we call *ego*. This ego creates a barrier, preventing us from seeing our oneness with God and oneness with one another. Growing in spirituality means growing beyond the ego.

A key point is that the quantum downward causation of choice is exerted discontinuously. If choice were continuous, a mathematical model or at least a computer algorithm could be constructed for it. As such, the outcome of the choice would be predictable, and its author would be redundant and could not be called God. Our ordinary waking state of consciousness, dominated by the ego, smoothes out the discontinuity by obscuring our freedom to choose, limiting the choice only to the known. To be aware that we choose freely is to jump beyond the ego, taking a discontinuous leap into the unknown—call it a quantum leap.

If you are having difficulty picturing a discontinuous quantum leap, a clarification by Niels Bohr can help. Bohr proposed a model of the atom in 1913. He suggested that electrons can move only in certain ways. Electrons go around the nucleus in continuous orbits. But when an electron jumps from one orbit to another, it moves in a very discontinuous way; it never goes through the intermediate space between the orbits. It disappears from one orbit and reappears in the other, causing energy quanta to be emitted or absorbed, depending on the direction of the jump. The jump is a *quantum leap*.

How does the cosmic, nonlocal quantum consciousness, God, identify with an individual, become individualized? Or, how does an individual experience his or her God-consciousness? How does continuity of the mundane world obscure discontinuity? Primarily via observership, and secondarily via conditioning.

Before observership, our God-consciousness is one and undivided from its possibilities. Observership implies a subject-object split, a split between the self that observes and the world that is observed. The world-experiencing subject or self is unitive and cosmic in the primary experience of a stimulus. In this primary experience, our God-consciousness chooses its response to the stimulus from the quantum possibilities with total creative freedom, subject only to the constraint of the laws of quantum dynamics governing the situation.

With additional experiences of the same stimulus, experiences that lead to learning, our ego responses become biased by past responses to the stimulus. This is what psychologists call *conditioning* (Mitchell and Goswami, 1992). Identifying with the conditioned pattern of stimulus responses (habits of character) and the history of past responses gives the subject/self an apparent local individuality, the ego. (For further details, see Goswami, 1993.)

When we operate from the ego and our individual patterns of conditioning, our experiences, being predictable, acquire an apparent causal continuity. As a result, we develop a greater sense of our personal self. We feel separate from our unitive whole self and from our God-consciousness. It is then that our intentions don't always produce the intended result.

THE QUESTION OF FREE WILL

The sum and substance of conditioning is that as consciousness progressively identifies with the ego, there is a corresponding loss of freedom. At the extreme of infinite conditioning, the loss of freedom is 100 percent. At that point, the only choice left to us, metaphorically speaking, is the choice between very familiar flavors of ice cream: chocolate or vanilla, a choice between conditioned alternatives. Not that we want to depreciate the value of even this much freedom, but obviously this is not real freedom. At this extreme, behaviorism holds; it is the so-called *correspondence principle* limit of the new science. (The *correspondence principle* in quantum theory was formulated by Bohr in 1923, according to which quantum and classical Newtonian theories tend to agree in certain situations, for example, in the macroscopic domain of reality. The conditions under which quantum physics and classical physics agree are called the *correspondence limit* or the *classical limit*.)

But do not fear. We never go that far in our conditioning. Even in our ego, we retain some freedom. A most important aspect of the freedom that we retain is the freedom to say "no" to conditioning, a freedom that allows us to be creative every once in a while.

There are experimental data that support this position. In the 1960s, neurophysiologists discovered the *P300 event-related potential* that suggested our conditioned nature. (In brief, a P300 ERP is a short—300 milliseconds—electrical wave in a person's electroencephalogram [EEG]. The P300 is used as an index of mental activity, a measure of how the brain waves discriminate between potentially important stimuli and non-important stimuli. The amplitude of the P300 wave increases with stimuli that are unpredictable, unlikely, or highly significant.)

Suppose that, as a demonstration of your free will, you declare your freedom to raise your right arm and then you proceed to do it. Guess what? By looking at an electroencephalograph attached to your brain, a neurophysiologist can easily predict from the appearance of the P300 wave that you are going to raise your arm. What kind of free will do you have if your decision can be predicted?

So then, is it the behaviorist who is right? Is there no free will for the ego? Maybe the mystics are right—the only free will is God's will, to which we must surrender. And then a paradox: how do we surrender to God's will if we are not free to surrender?

But again, do not fear. The neurophysiologist Benjamin Libet (1985) did an experiment that rescues a modicum of free will even for the ego. Libet asked his subjects to negate action as soon as they became aware of their being able to freely will to raise their arms. In that case, neurophysiologists would still predict from the P300 that they would raise their arms. But more often than not, Libet's subjects were able to resist their will and not raise their arms, demonstrating that they retained their free will to say "no" to the conditioned action of raising their arms.

EXPERIMENTAL EVIDENCE OF DISCONTINUITY

There are many situations in which analysis makes it unambiguous that electrons make quantum leaps quite routinely, not just situations in which atoms emit light as a result of these leaps. For example, there is the phenomenon of radioactivity, in which electrons sometimes come out from the nuclei of the radioactive atoms. Analysis shows that the electrons "penetrate an energy barrier." But how can an electron penetrate an energy barrier when it doesn't have enough energy to jump over it? Some physicists use the term "tunneling" to describe this phenomenon. The electron passes the energy barrier by making a quantum leap, without going through space to do its tunneling. Now it is on this side of the barrier; an instant later, it is on the other side, with a quantum jump.

But analysis is still just theory. Are there experiments that actually show that electrons are not continuously passing through an energy barrier, but really discontinuously quantum-jumping it? Yes. The same kind of "tunneling" phenomenon is found in certain transistors. In that case, experimenters have shown that the electrons make the transition from one side of the energy barrier to the other faster than the speed of light. Since, according to the experimentally verified theory of relativity, electrons cannot move faster than light in space, the electrons must be moving instantly without going through space. In other words, they are making a quantum leap.

In terms of possibility waves, the experimenter collapses the possibility wave of an electron on one side of the barrier—and then, immediately after, the electron is once again a possibility wave: one of its possible facets is that it is on the other side of the energy barrier. When our observation collapses the possibility wave on the other side, since no time elapses between the two observations, we must conclude that quantum collapse is discontinuous.

But it is a long way from a submicroscopic electron to a bulky human. How do we show that discontinuity is relevant for events pertaining to human consciousness to which everybody can relate? Are there indelible quantum leap signatures of the divine in macroscopic affairs of the world? Yes.

Is Creativity a Quantum Leap?

I hope the question of creativity being a quantum leap is not evoking images of creative people such as Newton, Michelangelo, and Martha Graham effortlessly jumping over great physical barriers. As you undoubtedly recognize, on the physical plane quantum effects tend to be smoothed out at the macro level. (See chapter 1.) We have to look at the mental plane, and that's where creativity is.

What is creativity? A little analysis will show you that work that we usually call creative consists of a discovery of new mental meaning—it involves a big change in how we process meaning.

Take the case of Einstein's relativity. When Einstein was a teenager, he came across a conflict between two theories of physics. On one hand, there was a theory by Isaac Newton; on the other hand there was a theory by James Maxwell—both great theories and both verified in their own right within the domain of their originators' intent. But the domains seemed to overlap and conflicts erupted in the domain of the overlap. Einstein worked ten long years on the problem, attempting to resolve the conflict; he made some progress, but a complete solution eluded him— until he woke up with a brilliant change of context for his entire framework of thinking. The context of the problem was two conflicting theories of physics, but the context of his solution was how we look at time.

Before Einstein, everyone thought that time was absolute, that everything happened in time and that clocks operated unaffected by movements. Wrong, said Einstein's creative insight. Instead, time is relative to motion. A moving clock, such as one carried on a spaceship, runs slower. This new context of looking at time resolved the conflict between Newton's theory and Maxwell's theory, and it enabled Einstein to develop a new mechanics from which came the wonderful idea of $E = mc^2$. This is an example of creativity. But was it discontinuous?

It had to be, because there was nothing manifest in anyone's thinking, either published or in scientific discussion, from which Einstein could have gotten the idea of moving clocks running slower. No algorithm could have given it to him. This is according to his statement, "I did not discover relativity by rational thinking alone."

To their credit, many scientists today agree with the idea that creative insights are quantum jumps in mental meaning and that they arrive discontinuously. This is partly because creativity research has solidly established, through many case studies, that creative insights in any field happen suddenly. How else would you explain the fact that one of the few established myths of science is about a creative event—Newton's discovery of gravity? I mean, of course, the apple story.

Cholera broke out in Cambridge in 1666, so Isaac Newton, a 23-year-old professor of physics, went to his mother's farm in Lincolnshire. There, while relaxing one morning under an apple tree in the garden, young Isaac saw an apple fall. And, wham! The idea of universal gravity, that all objects attract one another via the force of gravity, suddenly came to Newton.

Did it really happen like that? Some historians think that Newton's niece, when she was visiting France, started the story. But why did this story become part of the physics lore, when most of the physics community believed that science is done through trial and error—the scientific method—all logical and rational?

It's been said that mythology is the history of our souls. But when the traditional interpretations of the scientific discovery process as the result of continuous trial-and-error scientific method were not doing justice to the soul, guess what? A myth was created.

And of course, quantum leaps of creativity do not happen only in science. There is enormous evidence of discontinuous quantum leaps in the arts, music, literature, mathematics, and so on. You can find the evidence in many case histories compiled by creativity researchers. (Read, for example, Briggs, 1990.) You can also find the evidence in individual testimonies. Here are two samples:

Finally, two days ago, I succeeded, not on account of my painful efforts, but by the Grace of God. Like a sudden flash of lightning, the riddle happened to be solved. I myself cannot say what was the conducting thread which connected what I previously knew with what made my success possible. (mathematician Karl Fredrick Gauss, quoted in Hadamard, 1939, p. 15.)

Generally speaking, the germ of a future composition comes suddenly and unexpectedly.... It takes route with extraordinary force, shoots up through the earth, puts forth branches and leaves, and finally blossoms. (composer Tchaikovsky, quoted in Harman and Rheingold, 1984, p. 45.)

I think the best proof for the discontinuity of the quantum leaps of creativity is our own childhood experiences of learning new contexts of meaning. The philosopher Gregory Bateson classified learning in two ways. Learning 1 is learning within a given fixed context of meaning; for example, rote learning, memorization. But there is also learning 2, according to Bateson, involving a shift of the context. This one takes a quantum leap.

When I was three years old, I remember my mother teaching me numbers. At first, I was memorizing how to count up to 100. Not much fun, but I did it because my mother drilled me. She fixed the context. The numbers had no meaning for me. Then she was telling me about sets of two—two pencils, two cats—or sets of three—three rupees, three shirts. This went on for a while, and then one day, unexpectedly, I got it. The difference between two and three (and all other numbers) became clear to me. Implicitly, I had understood numbers within a new context—the set—although of course not in that language. And it was an extremely joyful experience. (Mind you, though, the concept of set was implicit, not explicit, in my consciousness when this experience occurred. In those days, sets were not introduced that early in our education.)

In the same vein, you may remember the experience of comprehending connected meaning for the first time when reading a story. Or the experience of comprehending what the purpose of algebra is. Or you may have had the experience of comprehending how individual notes, properly composed, make music come alive. Our childhood is full of the quantum leaping of such experiences.

Even dolphins are capable of taking quantum leaps of learning. Gregory Bateson (1980) tells the story of training a new dolphin under his guidance.

The animal went through a series of learning sessions. In each, whenever the dolphin did something that the trainer wanted repeated, he would blow a whistle. If the dolphin repeated her behavior, she would be rewarded with food. This is the usual training for showcase dolphins.

Bateson introduced the additional rule that the dolphin would never be rewarded for behavior already rewarded in a previous session. But in practice, the trainer could never maintain Bateson's rule, because the dolphin would be so upset about being wrong and not getting fish!

In the initial 14 sessions, the dolphin was just repeating the behavior previously rewarded and getting unearned fish if she was too upset. Once in a while, she was doing something new, but seemingly only by accident.

> However, between the fourteenth and fifteenth sessions, the [dolphin] appeared to be much excited, and when she came onstage for the fifteenth session she put on an elaborate performance including eight conspicuous pieces of behaviour of which four were entirely new—never before observed in this species of animal. From the animal's point of view, there is a jump, a discontinuity. (Bateson, 1980, p. 337)

TANGLED HIERARCHY

You may not have noticed, but there is another way that we can see a paradox in the observer effect. The observer chooses, out of the quantum possibilities presented by the object, the actual event of experience. But before the collapse of the possibilities, the observer himself or herself (his or her brain) consists of possibilities and is not manifest. So we can posit the paradox as a circularity: observer (brain) is needed for collapsing the quantum possibility wave of an object; but collapse is needed for manifesting the observer (brain). More succinctly, no collapse without an observer; but no observer without a collapse.

If we stay on one level, the material level, there is no solution to the paradox. The consciousness solution works only because we posit that consciousness collapses the possibility waves of both observer (brain) and the object from the transcendent reality of the ground of being that consciousness represents.

The artificial intelligence researcher Douglas Hofstadter (1980) gave us the clue for understanding what is occurring. Such circularities, he noted, are called *tangled hierarchies.* Most interesting is that self-reference, a subject-object split, emerges from such circularities.

Let's consider an example given by Hofstadter. Consider the *liar's paradox* expressed in the sentence, *I am a liar*. Notice the circularity: if I am a liar, then I am telling the truth, and if I am telling the truth, then I am lying, and so on *ad infinitum*. This is a tangled hierarchy because the causal efficacy does not lie entirely with either the subject or the predicate, but instead fluctuates unendingly between them. These infinite oscillations have made the sentence very special—the sentence *is speaking of itself, separate from the rest of the world of discourse.*

But this apparent separation of the self of the sentence and its world depends on our understanding the rules of English grammar and staying within them. The circularity of the sentence disappears for a child who will ask the speaker of the sentence, "Why are you a liar?" The child fails to appreciate the tangle and get caught up in it because the language rules are obscure to him or her. But once we know and abide by these language rules, we are looking at the sentence from inside and we cannot escape the tangle. Grammar, although the real cause, is implicit, transcending the sentence.

Similarly, in the observer effect, the reason it took us physicists a while to decipher the situation was because the choosing consciousness—God—is implicit, not explicit; transcendent, not immanent. The collapse is tangled-hierarchical, giving the appearance of self-reference or of the subject-object split. However, the observer-I, the apparent subject of the collapse, arises codependently with the object.

Whenever there is a collapse of a quantum possibility wave, there is a tangled hierarchy in its measurement. Along with nonlocality and discontinuity, tangled hierarchy is another indelible quantum signature of divine downward causation.

So the idea of tangled hierarchical quantum measurement is the final step that gives us a completely paradox-free solution to the quantum measurement problem that has puzzled physicists for decades. Additionally, this one idea helps us solve several very big mysteries of reality.

In the 1980s, I was talking with a Chilean physicist about the idea of consciousness collapsing the quantum possibility wave. He immediately raised the question, "At the moment of the 'hot' big bang creation of the universe, there were obviously no conscious observers around. So, pray tell, how did the universe collapse into actuality?" When I chuckled and showed him the solution (see chapter 7), he was mollified.

There is also the problem of the origin of life that haunts biologists even today. Apply the lessons of quantum measurement theory to that problem, and the solution springs out. (See chapters 7 and 8.)

The concept of the unconscious was introduced by Sigmund Freud in psychology. Since then the idea has been experimentally verified. Despite all of the recent successes of cognitive psychology, it is a fact that these scientists cannot explain how to distinguish between the unconscious and the conscious and how the subject-object split of conscious awareness arises. These problems are also solved using the idea of tangled hierarchical quantum measurement. (See chapter 6.) And in all these solutions to some of the most serious scientific research problems, we find indisputable evidence and support for the quantum God hypothesis.

Chapter 6

Downward Causation in Psychology:
The Distinction Between Conscious and Unconscious

B y all accounts, Sigmund Freud was an atheist. He ridiculed spiritual oceanic experiences as examples of infantile helplessness. He broke with Carl Jung, his most promising protégé, because Jung tended to take religion seriously. So why is the theory that Freud founded, psychoanalysis, held in such ridicule by a Nobel Prize-winning materialist, the physicist Richard Feynman? Here is what Feynman (1962) said:

> Psychoanalysis is not a science: it is at best a medical process, and perhaps even more like witch-doctoring. It has a theory as to what causes disease—lots of different "spirits," etc.

So Feynman seems to be saying that psychoanalysis has a lot of concepts, such as the unconscious, that smack of the witch doctor's "spirits." Materialists don't like the unconscious, because with materialism it is quite impossible to distinguish between the conscious and the unconscious. In 1962, when Feynman wrote the comments above, he also said, "Psychoanalysis has not been checked carefully by experiment."

But now the unconscious, the most important idea of psychoanalysis, has been thoroughly verified from quite a few different angles. And this has opened up another impossible question for the materialists.

So does the God hypothesis and downward causation help us distinguish between the unconscious and the conscious? You bet!

QUANTUM MEASUREMENT IN THE BRAIN AND THE UNCONSCIOUS-CONSCIOUS DISTINCTION

How do we perceive a stimulus that involves measuring it? How do we measure it? The crucial point is to recognize that in every event of perception and its quantum measurement, we not only measure the object we perceive, but also the state of the brain. Before we measure, the object is a wave of possibility, but so too is the stimulus that the brain receives from a possibility, a stimulus in possibility. And upon receiving such a stimulus, the brain too becomes a wave of possibility, a bundle of possible brain states. When we choose the state that actualizes the object we perceive, we also have to choose from among the possible brain states.

Face it. There is a paradox here. The brain (and the object/stimulus) remains in possibility until a choice among its possible states has been made. But without a brain, we cannot say that there is an observer, a subject's I (albeit in unitive consciousness) that is doing the choosing. This is a circularity that we call a *tangled hierarchy*, a concept that I introduced in the last chapter.

Simple hierarchy occurs when one level of a hierarchy causally controls the other(s). Figure 1-1 (page 17) depicts a simple hierarchy. To understand a *tangled hierarchy*, examine the Escher picture, "Drawing Hands" (figure 6-1). It is a tangled hierarchy because the left hand is drawing the right hand as the right hand is drawing the left hand. The causal control oscillates. (This is a version of the old question, "Which came first, the chicken or the egg?") The tangle can be seen clearly and also be resolved by "jumping out of the system," thinking outside the frame, realizing that neither hand is drawing the other—Escher is drawing them both. We cannot see the tangled hierarchy if we remain

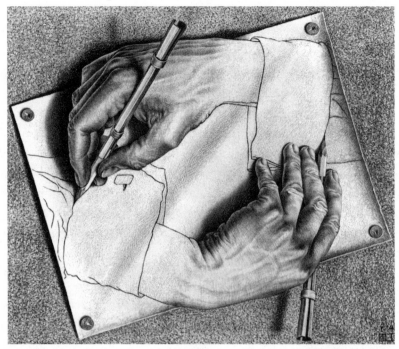

FIGURE 6-1. "Drawing Hands" by M.C. Escher. An example of tangled hierarchy.

within the infinitely oscillating system. Instead, we get stuck and think of ourselves as *separate from the rest of the world*. In this way, a tangled hierarchy gives the appearance of self-reference (Hofstadter, 1980).

So, quantum measurement involving the brain is tangled-hierarchical. The reward is that we gain the capacity for self-reference, the ability to see ourselves as a "self" experiencing the world as separate from us. The downside is that we don't realize that our separateness is illusory, arising from a tangled hierarchy in quantum measurement, a quantum collapse. Quantum measurement in the brain is special because of this tangled hierarchy involved in the passage from micro to macro in the neurophysiological processing of an external stimulus leading to perception.

Neurophysiologists try in vain to decipher the stages in which a stimulus is processed. Take an optical stimulus, for example. A photon from the object arrives at the retina of an eye and then travels along a nerve as an electrical stimulus to a brain center, etc, etc. To their credit, the neurophysiologists can do the analysis for a bit, but then everything gets jumbled up. The brain is too complex.

We can, however, recognize what is involved in the final reckoning. For a quantum measurement to have taken place in the brain, there must be a series of apparatuses that process and amplify the stimulus, taking it from the microscopic scale to the macroscopic scale. Somewhere in this passage from the micro level to the macro level, the tangled hierarchy is created because there is an infinite feedback loop that is impossible to decipher or break down into steps. We cannot follow the steps logically, but we can depict the end result—self-reference, as in figure 6-2. Finally, what we perceive is the object that sent the stimulus. We do not perceive the brain state, the brain representation of the object; instead, the unitive consciousness identifies with the brain state collapsed and memorized, and experiences itself as the subject of the collapsed object.

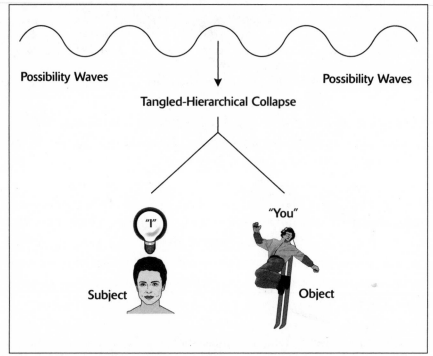

FIGURE 6-2. Tangled hierarchical quantum measurement of an object/stimulus in the brain produces not only the experience of the object, but also the experience of the subject in our consciousness.

Whenever there is such a tangled hierarchy in a quantum measurement situation, there is self-reference: the subject that perceives (senses) and the object that is perceived (sensed) arise codependently.

So what is the distinction between unconscious and conscious? Unconscious is when there is processing but no collapse. Possibility objects interact with other possibility objects in the unconscious, expanding in possibility. There is consciousness and processing, but no awareness. We call this unconscious processing, that is, processing without collapse, without subject-object awareness. And then there is the conscious, when there is collapse, when there is subject-object split awareness.

We can go through essentially the same analysis for a mental object of meaning. We recognize that the mental object cannot collapse of itself; there is no micro-macro division within the mind, no tangled hierarchical measurement-aid apparatus. But mental meaning can correlate with a physical object. When the physical object is collapsed, the correlated mental meaning collapses. In this way, the brain memory that results from a particular collapse event is not only a memory of the physical object, but associatively also a memory of the mental meaning. We can say that the brain has made a representation of the mental meaning.

Let's go back to Freud and his concept of the unconscious. Freud created some confusion with his language. What he called "unconscious," he should have called "unaware." When we recognize that consciousness is primary, we also realize that consciousness is always present. It is the ground of all being, so where would it go if there's an unconscious?

Freud's concept of the unconscious is actually much narrower than what quantum physics indicates. Quantum consciousness that collapses an original stimulus, an object—mental and/or physical—that we have never before encountered, is experienced in its full creative and unconditioned glory. The unconscious processing that precedes such a collapse event is also unrestricted, unlimited by conditioning. This is more like the concept of collective unconscious that Jung introduced in psychology. (This concept, which Jung later labeled the *objective psyche*, is the set of typical modes of feeling, thought, expression, and memory that seem to be innate to all human beings.) In contrast, what Freud originally meant by "unconscious" can be called our *personal unconscious*.

Let's examine the difference more clearly. Memories accumulate in the brain as we experience and learn about our stimuli. More and more, as our unconscious processes our memories, the tendency is to process every stimulus in terms of what we remember from previous experiences of stimuli. Soon we develop a habit pattern in which we use our mind to give meaning to our experiences, a pattern that we call our *character*. This character, plus the accumulation of our memory/history, is the quantum physics version of what psychologists call the *personal self* or the *ego*. The ego also subverts the tangled hierarchy of primary experiences into a simple hierarchy (one level causally controlling the other) in which the ego chooses within the context of its learned "programs."

Now to Freud's concept of the personal unconscious. Some of our experiences are traumatic, so much so that we are extremely reluctant to experience them ever again. However, they are acquired by our memory, just as are all experiences. So we develop an extreme resistance against retrieving any of those memories. Even in our conditioned ego, we retain the power of intention, of refusing to collapse a possibility. In this way, the possibilities we want to avoid experiencing, we are most often able to avoid. Unfortunately, unconscious processing of what we thus suppress continues without our control. So the repressed memory affects the overall processing of new possibilities, sometimes producing reactions that cause what we can call deviant, irrational, or mentally sick neurotic behavior. Freud's description was simpler but quite to the point: the unconscious id acts as a force to make us behave in ways that we cannot explain through a stream-of-consciousness analysis of our behavior. Our behavior has become irrational.

In materialist thinking, all is the play of the forces of upward causation. There is no room for the force of an unconscious id, which would cause havoc in the world of behavioral psychology, the psychology of conditioned behavior. So Freud's psychoanalysis is anathema, voodoo psychology to the materialist.

So although Freud was an atheist, his psychology of the uncon-scious, now called *depth psychology*, gives us indisputable evidence for downward causation or for what is its agency, God. The causal power of the unconscious id originates from the divine power of downward causation that we retain, albeit in a limited sense, even in our condi-tioned ego.

THE COLLECTIVE UNCONSCIOUS

Whereas Freud's discovery of the personal unconscious acknowledged a trickle of the potential power of downward causation, in Jung's con-ceptualization of the collective unconscious, that trickle became a mighty torrent. The collective unconscious holds our collective nonlo-cal memory, said Jung. Its movements, of which we are unaware, erupt in our awareness in the form of archetypal experiences in creativity and "big" dreams. (Jung used the term "big dream" to refer to a dream that contains universal significance, that is important because of the univer-sality of its images, archetypes.) They also precipitate events of syn-chronicity in which the archetypes of the collective unconscious show their "psychoid" nature: they causally affect not only events in the psy-che, but also events outside the psyche, in the physical reality itself.

The concept of synchronicity implies no less than the power of downward causation of consciousness (God) mediating between mat-ter and psyche. No wonder when somebody once asked Jung, "What do you think of God?" Jung promptly replied, "I don't think. I know." And Jung also said, "Sooner or later nuclear [quantum] physics and the psychology of the unconscious will draw closer together as both of them, independently of one another and from opposite directions, push forward into transcendental territory, the one with the concept of the atom, the other with that of the archetype" *(Aion, 1951)*.

A little clarification of terminology is needed here. What Jung called the *collective unconscious* is what we identify as the *unmanifested con-sciousness*, most of which belongs to the supramental domain. Jungian archetypes are the mental representations of the Platonic archetypes (supramental forms or ideas or patterns according to which all things

89

are constructed, which are understood by insight, as if by recollection, rather than by perception through the senses), which define movement in the supramental domain. From prehistoric time, human beings have intuited these archetypes and represented and labeled them; they are the gods and goddesses of our mythology.

Whereas Freud's vision of downward causation is myopic—it deals only with pathology—Jung's vision is far-reaching: it concerns the human potential, which, according to Jung, is "to make the unconscious conscious," to make the unmanifest manifest. For Jung, the human potential culminates when we have represented and integrated all the archetypes of our unconscious and actualized our Self. Then we are "individuated."

The new science agrees with Jung and accordingly chalks out an evolutionary pathway for humans striving toward individuation.

DIRECT EVIDENCE FOR UNCONSCIOUS PROCESSING

There is now a lot of direct evidence for the unconscious and for unconscious processing. The first piece of evidence is a striking phenomenon called *blindsight* (Humphrey, 1972). There are people who are cortically blind (experiencing a loss of vision because of an abnormality of the visual cortex of the brain), but who have vision processing in their hindbrains that is entirely unconscious. (The hindbrain is basically a continuation of the spinal cord; it receives incoming messages first and controls functions of the autonomic nervous system such as breathing, blood pressure, and heart rate.) In other words, a blindsight person can "see" via the hindbrain (unconsciously) and behave accordingly, but since this person is not seeing with his visual cortex (consciously), he would deny it. In a typical experiment, these seemingly blind people would be asked to travel in a straight line that contains an obstacle. The data show that the subjects would always go around the obstacles, but when the experimenters asked them why they deviated from their straight-line path, they would be puzzled and say, "I don't know." Clearly, they were processing or "seeing" the obstacles unconsciously but were unaware of them.

Like the processing by the hindbrain, processing by the right corti-cal hemisphere (the right brain) is also entirely unconscious. Experiments have been carried out with split-brain patients, whose left and right brains are surgically disconnected (that is, the main link, the corpus callosum, has been severed), except for the cross connections in the hindbrain centers involved with the processing of emotions and feelings. In one experiment, the experimenter projected the picture of a nude male model into the right brain hemisphere of a female subject in the midst of a sequence of geometrical patterns. The subject blushed, but when asked why, she couldn't explain. Seeing the nude picture and the feeling of embarrassment must have been processed unconsciously.

The best available data for unconscious processing, in this author's opinion, are collected in connection with near-death experiences. Some people after a cardiac arrest die clinically (as shown by a flat EEG read-ing), only to be revived a little later through the marvels of modern med-icine (Sabom, 1981). Some of these near-death survivors report having witnessed their own surgery, as if they were hovering over the opera-tion table. They are uncannily able to give specific details of their oper-ations that leave no doubt that they are telling the truth, however difficult it is to rationalize their autoscopic vision in their near-death experience. Well, they are not "seeing" with their local eyes, with sig-nals—that much is clear. Indeed, even blind people report such auto-scopic vision during their near-death comas (Ring and Cooper, 1995). These patients are "seeing" with their nonlocal, distant-viewing ability using the eyes of others involved with the surgery—doctors, nurses, etc. (Goswami, 1993). But this is only half of the surprise that the data present.

Try to understand how they can "see" even nonlocally while they are "dead," unconscious, and quite incapable of collapsing possibility waves. This is through unconscious processing, of course, which is like the people with blindsight, except that, unlike the latter, the near-death survivors have memories of what they processed while unconscious (Van Lommel et al., 2001). A chain of uncollapsed possibilities can col-lapse retroactively in time. This has been verified in the laboratory via the delayed choice experiment. (See chapter 7.) For the near-death survivor, the "delayed" collapse takes place at the moment the brain function returns, as noted by the EEG, precipitating a whole stream of collapses going backward in time.

The near-death data may be the most impressive, but the most important evidence of unconscious processing occurs in the phenomenon of creativity.

UNCONSCIOUS PROCESSING IN THE CREATIVE PROCESS

We discussed the discontinuous quantum leap of creativity earlier. It is important to recognize that the discontinuity of creativity is not an isolated event. If it were, a scientific study of it would be relatively fruitless because of a total lack of control. Fortunately, this is *not* the situation.

It is now well established that the creative process consists of four distinct stages (Wallas, 1926): preparation, unconscious processing, insight, and manifestation. The first one and the last are obvious: preparation is reading up and getting acquainted with what is already known; manifestation is capitalizing on the new idea, obtained as insight, by developing a product. These stages are both done more or less in a continuous fashion and with a lot of control. But the middle two processes are more mysterious. They are the analogs of the two stages of quantum dynamics: the spreading of the possibility wave and the discontinuous collapse.

As already discussed, unconscious processing is processing during which we are conscious but unaware. In creativity, unconscious processing accounts for the proliferation of the ambiguity of thought. Its analog is the spreading of the quantum possibility wave between measurements. Creative insight, of course, is sudden and discontinuous. As discussed in chapter 5, it is the analog of the electron's quantum leap from one orbit to another without going through the intervening space. An insight is a discontinuous quantum leap from one thought to another without thinking through the intermediate steps. Unconscious processing produces a multitude of possibilities; insight is the collapse of one gestalt of these possibilities (a new one of value) into actuality. (More on this in chapter 17.)

In this way, the creative process is an undeniable mixture of both continuity and discontinuity. The discontinuity we cannot control, but the continuity we can. And this makes creativity a scientifically tractable phenomenon.

THE GUIDING RULES OF THE NEW SCIENCE

I want to make an important point in passing. Every new paradigm of science brings along some modifications of the old standards of measurement. Before physics delved into the study of submicroscopic objects, the standard of observation was a strict "seeing is believing" or "show me." But submicroscopic objects like electrons cannot be seen in the old sense, with the naked eye. So we had to modify what constitutes seeing to include seeing via amplifying apparatuses. Next came quarks: they don't even exist in daylight, only in confinement. So now our concept of seeing in physics is further relaxed to mean seeing the indirect effects of quantum objects.

In creativity, the creator (the chooser of the insight) is the objective quantum consciousness. But the mental representation of the insight is made in the subjective ego, and through this subjectivity enters. Does this mean we cannot study creative insights scientifically? No, but we can't apply the criterion of strong objectivity—that events have to be independent of the observers or independent of the subjects. Instead we have to use weak objectivity—the events would have to be observer-invariant, more or less the same for different subjects, but independent of a particular subject. As physicist Bernard D'Espagnat (1983) has pointed out, quantum physics forces weak objectivity upon us already. And even experiments in cognitive/behavioral psychology cannot maintain a strict decorum of strong objectivity.

Above, we find one more relaxation of the protocol of the new science. In the old science, we demand total control and total power of prediction. In the new science, we are happy with partial control and therefore only limited power of prediction. But even with these new protocols, science can guide us adequately—and that guidance is the unique value of science.

Chapter 7

How God Creates the Universe and the Life in It

Who hasn't heard about the Big Bang, the explosive beginning of our universe according to modern cosmology? There is good empirical evidence for such an explosive beginning about 15 billion years ago in the form of a "fossil" remnant, a microwave background radiation that pervades our universe. Furthermore, the Big Bang fits well with the fact that our universe is expanding, as predicted by Einstein's theory of general relativity and observed by the astronomer Edwin Hubble.

When we attempt to incorporate the Big Bang within the theoretical framework that Einstein gave us for the large-scale structure of the universe—general relativity, in which gravity is seen as the curvature of space-time—the Big Bang appears to be a singular event. That explanation, in the 1960s, prompted an immediate response from theologians and astronomers alike: the Big Bang in its singularity must be a signature of the divine, of divine creation. Alas! It is not that simple.

QUANTUM COSMOLOGY

Thinking of the origin of the universe as a creation event is not entirely satisfactory. There is an anecdote about St. Augustine giving sermons on how God created heaven, earth, and all there is. One day, after his sermon, one of the backbenchers of the church heckled him: "Hey, Augustine, you always tell us about how God created heaven and earth. So tell me. What was God doing before he created heaven and earth?" It is said that, although Augustine was taken aback for a second, he recovered nicely and quipped, "He was creating hell for those who ask such questions."

The truth is that even with a singular beginning we can always ask, "What was before the singularity?" Also, the singularity is not a particularly desirable aspect of the theory of general relativity. This is because as the singularity is approached, such quantities of the theory as the energy density of the universe tend to expand to infinitely large values, which signifies that the theory itself must be questionable under those dire conditions.

Some cosmologists have addressed the issue of what preceded the Big Bang creation of the universe. Their ideas have led to many fashionable concepts, such as cosmic inflation. Among these theories, an idea of the physicist Stephen Hawking (1990) stands out: in the beginning, the cosmos must have consisted of quantum possibility. The universe must have been a superposition of many baby universes of possibility.

Hawking's reason for proposing such a quantum cosmology was to avoid the singular beginning in time. There is no beginning; there is only possibility. But now we must ask, "how does the superposition of possibilities become the actual universe that we inhabit?"

And there is a paradox that comes with a universe of possibilities and the question of how the possibilities can collapse into an actual event, the manifest universe. We know that it takes quantum consciousness acting through the form of a sentient observer to collapse quantum possibilities. It is impossible to imagine that there were conscious observers during the hot early days of the cosmos! So what gives?

Can the universe be here because of us, when we were not even there to greet it on the occasion of its Big Bang creation? Could we be putting the cart before the horse? Could it be that we are here because of the universe?

CHANCE AND NECESSITY?
OR ARE WE HERE BECAUSE OF THE UNIVERSE?

Many materialists think that we are here because of pure chance, some kind of cosmic accident. In materialist thinking, there is no meaning anywhere in the universe—and that includes us. "The more the universe seems comprehensible, the more it also seems pointless," said Nobel laureate physicist Steven Weinberg (1993).

Here's the materialist model of how the universe evolved. About a billion years after the Big Bang, statistical chance fluctuations cause galaxies to condense. Galaxies evolve, too, from the initial spherical cloud of gases to more of a disk shape, many having spiral arms. Then stars begin to condense, but these first-generation stars do not have all the elements needed to make life as we know it. In a few billion years, these first-generation stars go supernova, an explosion that leads to heavier elements. New second-generation stars condense out of the debris of supernova remnants along with planets. Some of these planets (such as the earth) have a solid core and evolve a suitable atmosphere, just what is needed for the evolution of life.

The play of chance continues, claims the materialist. Statistical fluctuations and the atmospheric energetics working together completely by chance make amino acids (the building blocks of proteins) and/or nucleotide molecules (the building blocks of DNA and other nucleic acid molecules). As you know, proteins and DNA are "living" molecules in some sense; they are the principal ingredients of a living cell and they have a tendency to survive maintaining their form. Now, according to materialist biologists, a new ingredient is added: the necessity for survival. Initially, there was only chance pushing the evolution of life on a planet like the earth; now it is chance and necessity, as in Darwin's theory. The rest is history.

This picture was at first supported by the famous Urey-Miller experiment in 1953, in which Stanley L. Miller and Harold C. Urey simulated conditions present on the early earth and tested for evidence of chemical evolution. That experiment demonstrated that amino acid molecules can spontaneously form in a watery solution of the basic atoms (carbon, hydrogen, nitrogen, oxygen) if the energetics of the early terrestrial atmosphere are suitably simulated. Later, the biologist Sol Spiegelman demonstrated in the laboratory that "living" macromolecules like RNA (ribonucleic acid) tend to preserve their form during chemical reactions, although ordinary molecules have no such tendency.

But problems remained. The huge production gap between the initial amino acids and the "living" macromolecule of proteins was never bridged. And theoretical calculations easily dispute the idea that chance can assemble a macromolecule like a protein from its basic ingredients, the amino acids; the probability is so small that it would take longer than the lifetime of the universe to do the assembly (Shapiro, 1986). Moreover, the probabilities cannot be expected to improve much even when we include survival necessity in the equation.

If evolution is not by chance and necessity, then is it by design? Is the universe purposive, made in such a way as to inevitably evolve sentience? Amazingly, today many astronomers and astrophysicists propound such an idea. It is called the *anthropic principle*.

THE ANTHROPIC PRINCIPLE

In its weak version (Barrow and Tipler, 1986), the anthropic principle states that the observed values of all physical and cosmological quantities do not arise with equal probability. Instead, they take on values restricted by the requirement that there exist sites where carbon-based life can evolve and by the requirement that the universe be old enough for this to have already happened.

The strong version of the anthropic principle is even more emphatic that there is a relationship between the universe and the life in it. It says (Barrow and Tipler, 1986):

The universe must have those properties which allow life to develop within it at some stage in its history.

Is the anthropic principle mere philosophy? No, it has much evidence behind it. It explains a lot of weird coincidences. I will give you a couple of examples.

You know the universe expands with time. If the force of gravity were even a tiny bit stronger, the expansion would rapidly change into collapse, so there would never be enough time for life to evolve. If gravity were a little weaker, the universe would go on expanding, but without any galaxies, stars, or planets to make a suitable environment for life. There is more. If the electrical force between electrons were even a little different, life as we know it would be impossible.

Such examples of the finely tuned universe can fill up pages. My favorite one involves the physics of the atomic nuclei—how three nuclei of helium atoms fuse together to make a nucleus of carbon, the all-important element for carbon-based life. The conventional wisdom regarding nuclear fusion reactions tells us that the probability of such a reaction should have been very low, too low to effectively generate a lot of carbon in the universe. But guess what? The conventional wisdom is wrong. The frequency with which the three helium nuclei vibrate as they come together exactly matches one of the natural frequencies of a vibrating carbon nucleus. The effect of such frequency matching is called a *resonance*, and it produces an enormous amplification of the reaction process, as when soldiers marching in unison on a bridge can destroy it.

How would the three helium nuclei know how to dance one of a select few dances that six protons and six neutrons of the carbon nucleus can perform? They could if there is a designer that designs both groups and designs the laws of whole nuclear physics to make such resonance happen.

The anthropic principle in both weak and strong versions suggests quite strongly that the universe is purposive, created by a designer with the purpose of creating life. Life is here, and by implication, we are here because of the universe. But an experiment of quantum physics suggests equally strongly that the universe is here because of us, the observers (see below).

Let me add one more idea and some comments on it. The materialist answer to the anthropic principle is the *multiverse theory*, which speculates that our universe is not unique, but is one among many. The idea is that if there are many, many universes, the odds are better that one of them would be fine-tuned enough to produce life. Well, this argument is not compelling for two reasons.

First, this is just a theory; even cosmologists admit that it is highly speculative. Let's wait for some verification. Needless to say, so far nobody even knows how to verify the existence of other universes!

Second, the argument presented by serious intelligent design theorists (Behe, 1996) is that life has an "irreducible complexity" built into it that makes it impossible to build life from matter via chance. Using quantum physics, I have made this argument foolproof, as you will see.

THE DELAYED-CHOICE EXPERIMENT

The physicist John Wheeler suggested an experiment to demonstrate that conscious choice is crucial in the shaping of manifest reality. This is called the delayed-choice experiment and has been duly verified in the laboratory (Hellmuth *et al.*, 1986).

In the delayed-choice experiment, we split a light beam into two beams—a *reflected* beam and a *transmitted* beam—of equal intensity, using a half-silvered mirror M_1 (figure 7-1). (A half-silvered mirror, aka a beam splitter, reflects 50 percent of the light and lets 50 percent of the light pass through.) The two beams of light are then brought together at a point of crossing P, using two regular mirrors, as shown in figure 7-1.

If we put detectors past the point of this crossing, as shown in the lower left of the figure, each detector will detect a photon (quantum of light) 50 percent of the time. Each detection event defines a localized path of the detected photon. The photon is showing its *particle* nature because its pathway is determined by the experimental arrangement.

But suppose we put a second half-silvered mirror M_2 at the crossing point P on the bottom right of the figure—what then? By splitting each of the two beams again into one reflected and one transmitted beam of equal intensity, *the second half-silvered mirror forces not one but*

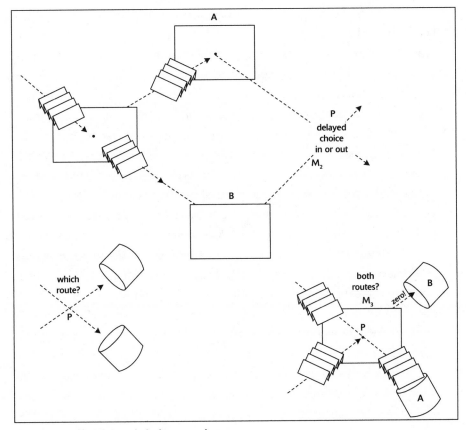

Figure 7-1. The delayed-choice experiment

two beams to operate on each side of P. Now two beams operating together interfere with one another like waves; hence, you have an opportunity to verify the *wave* nature of photons. Indeed, we can arrange the detectors in such a way that *the two waves interfere constructively at the detector site on one side* (A). At this location, the detector will certainly be activated. But at the detector location on the other side of *P*, the two waves destroy each other because they come together out of phase (B). Here the detector registers nothing and is never activated! How can this happen? To make sense of the experiment, we must assert that the photons are no longer traveling in localized paths as before; *they are traveling both paths until their detection, in possibility, as possibility waves.*

What you are witnessing is an experiment that demonstrates that light (and indeed, all quantum objects) is both waves and particles. As waves all quantum objects are transcendent waves of possibility; as particles they are immanent events of actuality.

But let's introduce another twist in the experiment. It takes the light a few nanoseconds (a nanosecond is a billionth of a second) to travel from M_1 to M_2. Suppose we insert the second half-silvered mirror at P within that time gap, after light has already been split at the first mirror. What happens now? If you think that the photons have already started on their designated path and will continue to show their particle nature, think again. The photons respond to even our "delayed" choice to put a second half-silvered mirror at P, and behave like waves and travel both paths.

On the other hand, if we were in the *middle of the wave detection experiment and there were already both mirrors M_1 and M_2* in their respective locations and we made the delayed choice to remove the mirror at P at the last nanosecond, what would happen? Again, the photons would respond even to our delayed choice and *travel one path or the other.*

Mind you, there is no paradox here, as soon as you reconcile in your mind that this is what it means when we say that light is a wave of possibility until we observe it! The entire path of the object stays in possibility until our observation manifests it *retroactively*. Yes, going backward in time.

You may have heard the story of the three baseball umpires comparing how they call their games and trying to one-up each other. The first umpire says, "I calls it like it is," as a Newtonian scientist would say. The second one is a little less egotistic; he may be a holist. "I calls it like I sees it," says he. But the third one is a quantum umpire at heart. He says, "There ain't nothing until I calls it."

So it is with the universe. There is nothing, no manifest universe, only possibilities, until we collapse it: until a sentient being appears in possibility in one of the possible branches and tangled-hierarchically observes. Then, the manifest universe.

The Delayed-Choice Experiment in the Macroworld

Many scientists are extremely impressed with the delayed-choice experiment, and it has helped change their attitudes toward the observer effect and the import of the anthropic principle. But there are still quite a few diehards reluctant to accept the message of the experiment because it applies to the microworld of things. "We will believe the potency of the observer when you demonstrate delayed-choice in the macroworld that we inhabit. Not before then." Well, the macroworld delayed choice experiment has been carried out, and successfully, by the physicist/parapsychologist Helmut Schmidt and collaborators (1993).

Originally, Schmidt had been researching psychokinesis, moving matter through conscious intentions, over many years with some measure of success. Some of these experiments involved the previously mentioned random number generators, which generate random sequences of zeroes and ones using random radioactive decay products.

His 1993 experiment was revolutionary because, with tremendous ingenuity, Schmidt was able to combine his psychokinetic experiments with random number generators and the idea of the delayed-choice experiment. In this experiment, the radioactive decay was detected by electronic counters, resulting in the computer generation of random number sequences that were then recorded on floppy disks. It was all carried out unseen by human eyes, months ahead of the time that the psychics came into the experiment. The computer even made a printout of the scores and, with such utmost care that no observer saw it, the printout was sealed and sent to an independent researcher, who left the seals intact.

A few months later, the independent researcher instructed the psychics to try to influence the random numbers generated in a specific direction, to produce either more zeros or more ones. The psychics tried to influence the random number sequence in the direction proposed by the independent researcher. Only after they had completed this stage did the independent researcher open the sealed envelope to check the printout to see if there was a deviation in the direction instructed.

A statistically significant effect was found. Somehow the psychics were able to influence even a macroscopic printout of data that, according to conventional wisdom, had been taken months ago. The conclusion is inescapable. "There ain't nothing" until an observer sees it: all objects remain in possibility, even macroscopic objects, until consciousness chooses from the possibilities and an event of collapse occurs. Then it all manifests, even retroactively.

BACK TO THE TANGLED HIERARCHY OF QUANTUM COLLAPSE

The lesson of the delayed-choice experiment is profound. The measurement problem of quantum cosmology—how the universe, looked upon as a wave of possibility, can ever be actualized, because obviously the harsh environment of a Big Bang creation excludes all observers—can now be resolved. The universe is a wave of possibility, a superposition of universes, and remains so until sentience evolves in one of its possible branches. When the first evolved sentient being observes, then the universe manifests retroactively, going backwards in time from that moment of collapse.

So it is true that we are here because of the universe and its purposive design, but it is also true that the universe is here because of us. There is circularity here, a breakdown of logic, which is crucial. Quantum collapse collapses not only the observed, but also the observer. This dependent co-arising crucially depends upon the circularity of the logical chain (see below).

There is also the important question, "What makes an observer?" We are used to thinking of ourselves, human beings, as the observers. Does the universe of possibilities wait in limbo all the way till the human observer comes onto the scene? This would confirm the Biblical idea that God created the immanent universe some 6,000 years ago.

However, this idea conflicts with the fossil evidence. But couldn't the fossils have been created retroactively, going backwards from the time of collapse 6,000 years ago, when Adam (in his God consciousness) first observed? Unfortunately for the Bible aficionado, this too contradicts the fossil data. The retroactive manifestation of the fossils would explain only fossils in the human lineage. The fossil data contain many other lineages, other kingdoms and phyla besides the animal kingdom and the phylum Chordata, to which humans belong.

I hope that the combined lessons of quantum cosmology, the anthropic principle, the delayed-choice experiment, and the fossil data are clear. Life itself, in the form of the first living cell, is the first observer.

What Is Life?

Biologists have no straightforward definition of life. In textbooks they gloss over it by giving the student a long list of properties shared by living systems. Quantum physics can rescue the biologists from their peculiar predicament of being unable to simply define the basics of what they study. If we say, "a living system has the capacity to observe," the biologists' consternation is over.

In truth, the Chilean biologist Humberto Maturana (1970) came close to giving us the above definition of life. He characterized life through the capacity for cognition. A little thought shows that cognition requires a cognizer, thinking requires a thinker, perception requires a perceiver—observer again.

The Observer and Circularity

Behold! The role of the observer in quantum measurement is clearly circular. The observer, the subject, chooses the manifest state of the collapsed object(s); but without the manifest collapsed objects that also include the observer, there is no experience of the subject. This circular logic of the dependent co-arising of the subject and object is called *tangled hierarchy*.

As mentioned earlier, the idea of tangled hierarchy and how it leads to self-reference or subject-object split has been explicated by the artificial intelligence researcher Doug Hofstadter (1980).

So how does self-reference arise in the living cell? Via tangled-hierarchical quantum measurement. Is there such a quantum measurement apparatus in the living cell? Yes—and this is the subject of the next chapter.

Chapter 8

The Design, the Designer, and the Blueprints for Design

The establishment biologist is stuck and reacts in only one way to any talk of design. Design to him raises the specter of a creator God, as in the Biblical Book of Genesis. In this way, design talk is not politically correct for a biologist.

But never fear! The application of quantum physics to the situation of life's origin readily shows that the Genesis model of God and creation is wrong; it is too simpleminded and linear. In the current creationism-evolutionism debate, even journalists without any science background thwart creationists by asking, "Who created the creator?" But the quantum God's creation of life and the universe is done in one fell swoop: the whole material universe waits in possibility until the first life is intuited and the self-referential quantum measurement circuit is completed. (The physicist John Wheeler called this the completion of the "meaning circuit." It is amazing how close Wheeler came to the ideas explored here.) The causal circularity forever rids us of the question, "Who created the creator?"

So there is no danger of succumbing to the literal Biblical creationists' ideas here. There is intelligent design, yes; evolutionism, yes; creationism, no. And we can relax.

I mentioned the Urey-Miller experiment in the last chapter. Its success in making amino acid from the basic atomic ingredients has inspired many a biologist to spend years in the vain hope of manufacturing life in the laboratory. The idea was to produce life through step-by-step chemical reactions, more and more complex molecules eventually ending up with a living cell. But this has not happened. And it never will. So right now, biologists are content to theorize how such step-by-processes might have come about; but none of these theories are convincing and there is no consensus.

The situation changes drastically when we put downward causation into the equation of life—God's creativity. The truth is that for the quantum creation of life, God does not need to follow this human-conceived step-by-step process. This scenario is the conception of the Newtonian mind that cannot see beyond continuity. God is quantum consciousness; God works with creativity, for which the operative phrase is *discontinuous quantum leap*. "Let there be life, and there was life."

However, the God, quantum consciousness, that we have rediscovered is lawful in our conceptualization up to a point. Even God's creative acts have to follow the usual process that creativity researchers have codified. And even a creator God, quantum consciousness, has to deal with the probability question: it is a fact that, as pointed out by many biologists, the creation of life is a very low-probability event. The way to deal with small probability, judging from our own creative experiences, is to have the benefit of a blueprint.

A DESIGNER NEEDS A BLUEPRINT: MORPHOGENETIC FIELDS

It is said in Genesis that God created man (woman) in His (Her) own image, signifying that we can extrapolate some of God's power by examining our own. It's even easier when it comes to creativity. In creativity, we are using God's power, God being our own creative agent of downward causation. So how do *we* create?

Consider an architect building a house. He will start with an idea. At the second stage, the architect will make a blueprint of the idea. Only then will he engage in actually creating the physical house, beginning with the physical components.

Quantum consciousness, God, follows the same procedure in creating life. He (She) starts with the possibility of an idea that belongs to the supramental domain. The blueprints of life belong to a compartment of possibilities called the *vital energy body*. Concomitantly, God starts processing (unconsciously, of course) the material possibilities for a physical representation of the vital possibilities (figure 8-1).

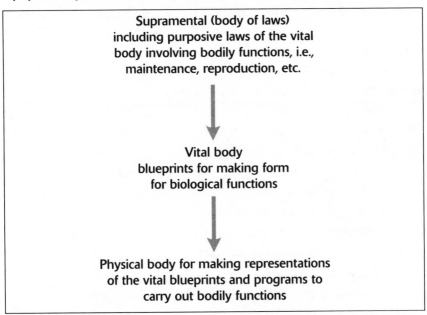

FIGURE 8-1. How a supramental context of living is represented in the physical via the intermediary of the vital blueprint—the morphogenetic field.

In the 19th century and even in the early 20th century, the vital body was considered an essential part of biological thinking. For example, the philosopher Henri Bergson saw life as an expression of *élan vital* or vital essence (energy?), that special feel of life from inside. Bergson's philosophy was quite popular among biologists. But the situation changed drastically after the discovery in the 1950s of molecular biology. The picture of a cell, containing DNA for replication and proteins for various functions of maintenance, seems to have all the ingredients to explain biological functioning. This and Darwin's theory of evolution, packaged in a new synthesis called neo-Darwinism, became the new paradigm of biology. The concept of vital energy was considered extra baggage and abandoned. It smacks of dualism anyway, which no scientist likes. Good riddance—or so the thought went!

But by 1960, biologists like Conrad Waddington (1957) were already pointing to a cloud on the horizon: the problem of biological form-making or, formally, *morphogenesis*: how a form like an organ is created from a single-celled embryo. A cell makes more of itself by cell division, making an exact replica with exactly the same DNA. But then why does a liver cell of a grown body behave so differently from a brain cell? How do the cells belonging to different organs get differentiated?

Obviously, the cells in different organs must be making different proteins by activating different sets of genes. This is called *cell differentiation*. Different programs activate the genes of the cells belonging to different organs. The source of these programs is called *morphogenetic field*, or so it is speculated.

But where are they, these morphogenetic fields? By the 1980s, a genetic or even epigenetic origin of these programs did not seem promising. The situation remains the same even today. Says the biologist Richard Lewontin (2000) about genetic models of morphogenesis:

> The processes of differentiation of an unspecialized cell into a mature specialized form are not well understood. The reason this is a deep problem for biological explanation is that cell differentiation lies at the basis of all development [of adult form from the embryo]. . . . It is very well to say that certain genes come to be transcribed in certain cells under the influence of the transcription of certain other genes, but the real question of the generation of form is how the cell "knows" where it is in the embryo.

I hope you can see the play of nonlocality here. At least one biologist, Rupert Sheldrake, saw this already in the early 1980s. Sheldrake (1981) wrote a book called *A New Science of Life* in which he proposed that the sources, the blueprints of the genetic programs of cell differentiation, the morphogenetic fields, are nonlocal, and that as such they can only be nonphysical.

Sheldrake analogically thought that these morphogenetic fields communicated their instructions to the cell as a radio transmitter does to a receiver, through a kind of resonance: frequency matching but without the local electromagnetic waves to communicate. Sheldrake called this mechanism the *morphic resonance*, that is, the resonant making of form.

Sheldrake's picture is dualistic, no doubt; a nonphysical morphogenetic field, albeit nonlocal, cannot interact or "resonate" with a physical cell without a mediator that correlates the two. Sheldrake at the time was reluctant to introduce the concept of a programmer (designer) that uses the blueprints to transcribe the programs of form into the picture.

Quantum thinking can do this without dualism. The blueprints (the morphogenetic fields), the programmable genes, and the form they create remain in potentia until God—quantum consciousness—makes a match (as in a resonance) and collapses actuality (Goswami, 1997b).

And now we can see clearly what this actuality is. The physical actuality is the form, the organ—this the biologist acknowledges and everybody can verify. But there is also the manifest morphogenetic field in awareness within the psyche of all living beings. This internal awareness is the feeling of being alive that Bergson called *élan vital*.

So let's connect this to an age-old terminology and unabashedly revive the term *vital body* as the reservoir of the morphogenetic fields. The movement of the vital body is naturally called the *vital energy*; this is what we feel whenever we actualize a functioning biological organ and its program.

Note also that the form, the organ, is initially made in God-consciousness, but when the form-making or creation is over and we begin using the forms or organs, our experience of feeling alive reflects more and more the effect of conditioning, of conditioned continuity.

BACK TO HOW GOD CREATES LIFE

Go back to the beginning of life. The probability of synthesizing a living cell's basic components in the laboratory—the protein and the

DNA—individually from amino acids and nucleotides, respectively, is minuscule (Shapiro, 1987). There is also a circularity here: the components of DNA—the genes—have the code for the assembly of amino acids into proteins. But proteins are required to make the DNA. We also know that finding a proper theory of how the DNA, the protein, and so on are assembled together in a cell, starting from the basic available nonliving ingredients, has eluded biologists so far. That such a theory will ever be developed and achieve consensus agreement is also highly unlikely.

But do we need a continuous process from the beginning to the end product? Let's invoke the discontinuity of creativity to complete the tangled-hierarchical quantum measurement system in the first living cell. The designer, quantum consciousness—God—recognizes the protein-DNA combination in possibility, albeit of small probability, because He (She) knows the purpose: self-reproduction and self-maintenance. God has the possibility blueprint of a living cell for guidance. The blueprint codifies the knowledge that to create a self-referential living cell, one needs a replicator system (DNA), maintenance managers (proteins), communicators between DNA and protein (RNA), cytoplasm for mobility, and a cell wall for confinement.

But in the actual physical production of the living cell from the quantum possibilities of the microscopic ingredients (amino acids, nucleotides, lipids, etc.), in the transition from micro (possibility) to macro (possibility) and then to macro (actuality), there is discontinuity; there has to be. The discontinuity arises from the fact that, short of the actualized living cell, no intermediate macro state of microscopic ingredients satisfies the requirements of a tangled hierarchy. As discussed earlier, the tangled hierarchy and discontinuity are the necessary price for self-reference or the subject-object split, and only God's creativity can resolve this paradox. This involves a high level of creativity able to quantum leap the usual continuity of a mechanical means of assembly; this requires an intelligent designer.

God makes humanity in His (Her) own image. The mathematician John von Neumann (1966), working with what are called *cellular automata* (little bits of programmed stuff), figured out theoretically the roles of the replicator and the maintainer systems in a truly reproductive system even before Francis Crick and James Watson discovered these roles in the laboratory.

But Von Neumann missed the self-reference of life, so his model is compatible with materialism; he did not see the necessity of invoking God and the quantum leap in the original production of life. Because we have recognized the importance of self-reference, we are figuring out how God must have created the first life, that only God-consciousness and a discontinuous quantum leap can actually create life from nonlife!

Holistic biologists such as Humberto Maturana and Francisco Varela see the materialist's predicament and propose holism, that the whole is greater than its parts: life is an emergent property of a living cell that cannot be reduced to its parts. But in view of what we know about how simple systems make up complex systems, such as atoms making up molecules, with no irreducibility there (since we know that molecules can be reduced to atoms and their interactions), the holists' claim sounds preposterous.

The biologist Michael Behe (1996) intuits much the same thing as the holist when he introduces the concept of *irreducible complexity*: biological features that are so complex that only an intelligent super-designer, God, can design them. Better language, perhaps, but no more plausible.

But now we understand what the holists and Behe are trying to say. It is the tangled hierarchy of life's design that is irreducible, that is irreducibly complex.

The magic of life comes from three things: 1) God's creative downward causation, creating 2) the tangled hierarchy in the organization of the living cell, giving us the self-referential quantum collapse in the cell, using the help of 3) the vital blueprint, of which the cell makes physical representation. The magic in the creation of life is the quantum leap of creativity; it can neither be synthesized by nor be reduced to bit-by-bit continuous evolution by mathematics, mechanical computation, or biochemical reactions in the laboratory.

People say that life is a miracle; this is not a cliché. Life is literally God's creative miracle gift to us.

THE PROOF OF THE PUDDING IS IN THE EATING

The bottom line is this: is there a tangible verifiable output we garner from this creativity theory of the origin of life? If there is a pudding from all the cooking (theorizing), then we can eat it and our doubt about the authenticity of the cooking will disappear. Similarly, we need experimental data to verify the validity of the theory. Otherwise it is all hocus-pocus, pseudoscience.

The materialist model is unabashedly devoid of purpose. Life has no purpose, life's origin has no purpose, and evolution has no purpose. Indeed biologists have even replaced *teleology*—purposivenes—with *teleonomy*—appearance of purposiveness.

There is one unsolvable mystery in this kind of thinking: why does evolution drive biological systems to more and more complexity? According to the biological establishment, evolution proceeds through the Darwinian mechanisms of chance variation and natural selection. Chance obviously can work both ways, toward complexity or toward simplicity. And natural selection favors fecundity, the ability to produce offspring with profusion, not complexity.

But if God created and evolved life, God did it for a purpose—for the purpose of seeing His (Her) possibilities manifest in physical representations. In this picture, the evolution of life is the evolution of more and complex representations of the blueprints of life contained in God's vital body, to see the possibilities of the vital body find better and better expressions.

Sometimes we call this drive of life to evolve in the direction of increasing complexity the *biological arrow of time*. And we can explain it only with a theory in which God is the creator of the original life and the causal driver behind the evolution of life (by virtue of subsequent quantum leaps—see chapter 9).

Incidentally, Darwin's chance and necessity give us a theory of *species adaptation* (how a species adapts to a changing environment), but not of evolution (how one species evolves into another), and especially not of macro *evolution* (evolution involving big changes in traits). The famous fossil gaps between species lineages are well known. If

Darwinian mechanisms—slow and continuous—held for all evolution, there would be no fossil gaps.

Behe (1996) makes the same point about the inadequacy of the neo-Darwinian model of evolution by considering the biochemistry of various "little steps" of evolution through Darwinian chance and necessity. And Behe reaches the same conclusion. Can Darwinian chance and necessity account for all the steps of evolution? No. Is the evolution of life also the product of a purposive designer? Yes.

To this we add: God is a creative designer who guides evolution through creative quantum leaps. The fossil gaps are evidence of the discontinuity of these quantum leaps. I have presented all these ideas in detail elsewhere (Goswami, 2008). In the next chapter, I give you a glimpse.

Chapter 9

What Do Those Fossil Gaps Prove?

Everybody knows about the fossil gaps—the apparent lack of geological evidence of transitions between distinct higher forms of life. Contrary to the expectations of a great number of biologists since Darwin, the fossil gaps have not yet filled up with thousands upon thousands of predicted intermediate life forms. The vast majority of the gaps are real. So what do they signify? What do they prove?

The neo-Darwinists—and the majority of biologists fall into this category, still insist that the gaps mean nothing. They are sold on a *promissory evolutionism*—the idea that eventually the gaps will fill up.

The most vocal public opponents of this view are followers of the Biblical Genesis creation or creationism, the idea that God created life literally as it says in Genesis, all at once. According to them, there is no evolution. Fossils mean nothing significant and the fossil gaps are the living (or should we say "dead"?) proof of that.

According to creationism, there cannot be any intermediates whatsoever. So today, biologists tout the few intermediates that are found to fill in the fossil data as evidence for evolution as well as for Darwinism. This is highly misleading. It is true that the existence of intermediates between two fossil lineages (as between reptiles and birds) refutes creationism and proves evolutionism, but evolutionism is not synonymous with Darwinism, which would require thousands upon thousands of such intermediates to verify it.

A slightly less radical group than either of the two above subscribes to a philosophy called *intelligent design*. Like creationists, they (at least most of them) believe (unnecessarily) that the fossil gaps mean no evolution ever. According to them, species do not change much and an intelligent designer created them all at once. Implicitly, the intelligent designer is assumed to be God, but no reference is made to the Bible.

It is easy to criticize the creationists and the intelligent design theorists. The Biblical account of the creation of the world and the life in it, if taken literally, is just plain wrong; the geological and radioactive dating evidence for the age of the earth is conclusive against it. The intelligent design proponents are also wrong in part. There is much evidence that species evolve from older species: we have too much common with monkeys, they have too much in common with mammals lower on the evolutionary ladder, and so forth. If you look at the early development of the embryo of a "higher" species, you will find that the stages resemble those of the development of lower creatures of an earlier era. The Darwinists got this one right! Later species evolve from earlier ancestors; there is no doubt about it.

But neo-Darwinists are dead wrong when they say that there is no meaning and purpose to evolution, that there is no play of intelligence in the design of life and how it evolves, and that there are no "lower" and "higher" creatures. Their insistence that evolution is a material process of blind chance and survival necessity is myopic. As Abraham Maslow said, "I suppose it is tempting, if the only tool you have is a hammer, to treat everything as if it were a nail." Neo-Darwinists are materialists; the hammer they have is the idea that everything is made of matter via upward causation, and that all life is the play of genes that carry hereditary information. There is no scope in such a philosophy to talk about meaning or purpose or intelligent design, except for any value that these ideas may have for survival of a species.

Let's take the case of meaning. For meaning to evolve as an adaptive survival value, matter must be able to process meaning. But in grammar, there is a category difference between syntax and meaning. The symbol processing by matter in the form of a computer is akin to processing syntax; so the idea of meaning processing by matter has always been a little suspect (Searle, 1980, 1994). And recent research (Penrose, 1989) has confirmed that computers and thus matter can never process meaning. (See Chapter 12.) How can nature select a quality from matter that matter cannot process?

This shortcoming, to explain intelligent qualities as evolutionary adaptation, becomes even more obvious when we ask, "How does our ability to discover scientific laws arise?" Such a discovery has survival value; that is not the question here. The question is "Can the knowledge of scientific laws be coded in matter somehow? Can they arise from the random motion of matter somehow?" Attempts to prove that this is the case have had no success whatsoever.

The question of how consciousness can evolve in matter is another case in point. "Can matter codify consciousness?" is the hard question. How can interacting objects ever produce a subject-object split awareness? If material interactions can never produce consciousness, to think of consciousness as an adaptive evolved value does not make any sense.

So intelligent design aficionados have got this one right—or have they?

Not quite. The conclusive scientific proof that there is purpose in God's creation is that there is a biological arrow of time. By looking at the fossil data, you can tell the direction of time—that time has gone from the past, from when the fossil data show only relatively simple life forms, to later times, from when they show more and more complexity of life forms. And only the most recent fossil data show us humans, the most complex of living creatures. So the purpose of evolution is to create complexity, and time's biological arrow moves from simplicity to complexity of living organisms.

All creationists and most intelligent design theorists deny evolution, and they justify their denial because an evolution in complexity is against the entropy arrow of time and is seemingly in violation of the entropy law—entropy always increases. These theorists, by denying evolution, are overlooking one of the best pieces of evidence for the existence of God. Of course, evolutionists miss the purpose and design in life.

So what do the fossil gaps signify? Apart from the slow tempo of evolution that Darwin suggested and neo-Darwinists agree on, there is also a fast tempo of evolution—so fast that there isn't time for the formation of fossils. This fast tempo is what produces the fossil gaps. In other words, evolution is like punctuated prose; there are abrupt and discontinuous punctuation marks within the otherwise continuous prose (Eldredge and Gould, 1972). The proponents of this idea are called *punctuationists*.

A class of biologists called *developmental biologists* (or organismic theorists who emphasize the role of the organism) has offered de facto support for this idea of a second tempo. This is because they believe that significant evolution at the macro level must involve the development of a novel organ. But a complex organ cannot evolve piece by piece. A little piece of an eye is useless for seeing. So such "macroevolution" must be discontinuous, requiring a fast tempo. But because there has never been any plausible suggestion of a mechanism for a fast tempo, the idea has not found general acceptance in the biological community. Scientists don't like living in an explanatory vacuum. If no theory of fast tempo is available, let's proclaim that Darwin's covers all evolution and explains away the fossil gaps!

There are biologists who point out another important piece of data that also suggests discontinuity. Before all great creative evolutionary epochs of macroevolution, there always occurs some kind of catastrophe leading to a massive extinction of biological species (Ager, 1981). These catastrophes clear up the biological landscape for a new evolution of species. And since the new evolved species have no need to compete for survival, another pet idea of Darwinism goes down the drain.

So here's what I intuit. Fossil data are some of the best proof of the existence of God and of God's creativity. Creativity occurs through quantum leaps, taking no time. I submit that here is the new mechanism for the fast tempo of evolution! I will show below that this theory integrates the thinking of everyone: of the intelligent design theorists, because the design arrived at via creativity is obviously intelligent and also because the designer is God; and of the developmental biologists, because indeed it takes creativity, one giant leap to "see" all the right possibilities for making a new organ and then making it. This theory satisfies the catastrophe thinkers, because death is part of creativity, destruction before creation. The destruction is also needed to open ground for the play of the new creations. The appropriate metaphor for God in this aspect of creativity is what Hindus would call Siva's dance (Figure 9-1). God in this special aspect of Siva, the king of the dancers, is first a destroyer and then a creator. The idea of creative evolution should even please the few open-minded Darwinists: Darwin's slow mechanism is the conditioned limit of God's creative downward causation—call it *situational creativity*.

FIGURE 9-1. Siva's dance—destruction before creation.

UNCONSCIOUS PROCESSING

The biggest problem of biological macroevolution is that such a giant step requires so many changes at the genetic level, so many mutations or variations! For example, the development of an eye from scratch requires literally thousands and thousands of new genes. But each gene mutation or variation, according to neo-Darwinism, is selected individually. The likelihood of its being individually beneficial (contributing some adaptive macro-level function to the organism) is quite small; in fact, gene variations are often just the opposite—downright harmful to the survival of the organism. So chances are high that individual selection would eliminate most gene variations. Considering this, it is easy to see that it would have to take a very long time—much longer than the geological time scale over which evolution occurs—to accumulate the many beneficial gene variations necessary for macroevolution.

However, this situation of biological macroevolution is saved by the idea of *unconscious* processing, part and parcel of the creative process. It is *conscious* processing that takes too much time, being guided by trial and error. But in quantum thinking, the gene variations are quantum possibilities anyway (Elsasser, 1981). Biologists using classical thinking assume that the quantum gene variations would collapse without any help from consciousness. But we know better: quantum collapse requires consciousness and its power of downward causation. And any gene that is not expressed in creating a macroscopic trait remains uncollapsed, even from one generation to the next. Consciousness does not collapse the unexpressed genetic variations—quantum possibilities all—until a whole configuration of them will make a new organ when expressed. Consciousness waits for the right moment, as we do in our own creative process.

What is crucial is that consciousness has the vital blueprint of the organ unconsciously giving it a rough guideline of what to process. When there is a match, a match that Rupert Sheldrake calls *morphic resonance*, a quantum leap takes place all at once and consciousness makes a physical (organ) representation of the morphogenetic blueprint expressing all the necessary uncollapsed genes at once (Goswami,

1997a, 1997b). There is *no fossil record for intermediate stages, because there are no intermediate stages!*

In this way you see clearly that the fossil gaps are evidence of biological creativity, of quantum leaps in evolution. And as such they provide us with the most spectacular evidence for God (as quantum consciousness) and God's creativity.

How about the occasional intermediate that shows up in nature? The morphogenetic blueprints are vital representations of archetypal functions. Sometimes in the journey of creative discovery, two archetypes become involved and their physical representations simultaneously give rise to an intermediate.

One question still needs to be addressed. Human creativity consists of the individual creatively taking a quantum leap to God (quantum consciousness), making the creative quantum collapse possible. Clearly, the individual has a role to play. What is the role of the individual organism in biological creativity? We will return to this question later.

SYNCHRONICITY

There is now consensus that the dinosaur extinction some 65 million years ago was brought about by a large meteor shower. This made room for the very important explosive evolution of the mammals, who were already on the scene but not as major players, which eventually led to the evolution of the human being.

So did the evolution of humans on earth occur through pure meaningless chance? If that is so, then how can we uphold the purposiveness of biological evolution, when clearly God's purposiveness needed the help of a chance event?

There is no contradiction with the scenario of biological creativity and purposiveness. Chance contingencies are often very important in the history of a creative act, except that we see them as components of synchronistic events.

Take, for example, the case of Alexander Fleming's discovery of penicillin. While Fleming was on vacation, a mycologist on the floor below his lab happened to isolate a strong strain of the penicillin mold that became airborne and found its way to a petri dish upstairs in Fleming's lab. An unusual cold spell for that time of the year helped the mold spores to grow while preventing bacterial growth. And then the temperature rose and bacteria immediately grew everywhere except in the petri dish. So a quantum leap occurred in Fleming's mind in the form of the question: What is in the petri dish that prevents bacteria from growing?

Similarly, an event outside in the material arena (the meteor shower) and one inside the biological arena (the act of biological creativity) occurred simultaneously, and meaning and purpose emerged in the evolution of many new mammals. This kind of coincidence of events is what the psychologist Carl Jung (1971) calls *synchronicity*.

In fact, as catastrophe theorists point out, these events of synchronicity are important because they open up the evolutionary landscape for the newly created macro organism. They also create a sense of survival urgency for evolution in the organisms that survive the catastrophe. A sudden change of environment requires an equally sudden evolutionary jump. There is no time for waiting for slow Darwinian evolution to bring adaptation.

THE ROLE OF THE ORGANISM

Now we can see the role of the organism in biological creativity that is responsible for the fast tempo of biological evolution. In neo-Darwinism, the organism has no role to play. This is bitterly disputed by organismic biologists, who maintain that the development of the organism, in fact the organism itself, must have a role to play.

In the above scenario showing how the quantum leap takes place, it is clear that development (of an organ) does play a crucial role. We can enunciate the role of the organism as well when we take account of the catastrophes that precede quantum evolution.

All creative people know that human creativity requires a motivation and an urgent demand, usually a burning question. From the point of view of the whole quantum consciousness or God, there is the motivation of purposive evolution (also, see later). When an environmental catastrophe occurs, this evolutionary motivation percolates through to the individual organisms in a hurry because it coincides with survival necessity.

I further suspect that biological organisms have nonlocal connections through the vital arena, the morphogenetic fields. Because of the dominance of the mind, this vital nonlocality is somewhat obscure for us humans. But the rest of the biological world, being nonmental or at least largely so, is not limited that way. So this nonlocal connection through the vital body acts as a species consciousness (a generalized species ego). I think it is this species consciousness that intends evolution in response to rapid environmental changes, and quantum consciousness/God responds to this evolutionary call.

CONNECTION WITH NEO-DARWINISM

In between the quantum leaps of quantum evolution, what happens? It is easy to see that the slow Darwinian mechanism is now enough to cope with slow environmental changes. Gradually, this builds up a gene pool of already environmentally adapted genes for the entire species, a pool that now can meet the adaptive needs of periodic environmental changes without having to develop new genes.

Note also that the creative leaps express a whole range of new genes. In some combination, these genes make specific organs. But a gene can be used and is used in more than one combination and in more than one context. In this way, you can easily see that the creative leaps of evolution also contribute to the cumulative buildup of the gene pool.

In human creativity, the ability to adapt to societal needs by inventing new combinations of old ideas is called situational creativity, as opposed to the fundamental creativity of discovery (Goswami, 1999). Thus the Darwinian mode of evolution can be seen as a special case of creative evolution involving situational creativity.

A good example is the famous case of the peppered moth around London, Birmingham, and other large industrial centers that underwent a change in color in the mid-19th century from speckled brown to black because of air pollution. The "black gene" was already in the gene pool.

The individual moths that were born with this "black gene" had an advantage over the speckled brown moths because their color camouflaged them better against trees blackened by soot. Hence, more of the black moths survived predator birds and fewer of the speckled brown moths did. So, rapidly, natural selection wiped out many of the formerly predominant speckled brown moths and favored the black moths.

Finally, as Stephen Gould and others (intelligent design theorists included) have noted, the fossil data also show vast epochs of virtual stasis in the evolutionary history of all species. This corresponds to the limit of conditioned existence when no creativity, situational or fundamental, was needed.

THE BIOLOGICAL ARROW OF TIME AND THE FUTURE OF EVOLUTION

As mentioned above, there is a clear biological arrow of time: biological organisms evolve from simplicity to complexity. What defines "complexity" should also be clear from our account of creative evolution. Complexity consists of new organs, which are either more sophisticated expressions of previously expressed biological functions or expressions for entirely new functions previously not represented in the physical.

Neo-Darwinism cannot explain an arrow of time. Both of its steps, production of chance variation and natural selection, are no more likely to favor complexity over simplicity than simplicity over complexity. Chance is, of course, another word for random: so chance variation can lead to simpler designs or more complex designs. Natural selection also, in the final reckoning, selects only according to fecundity, the capacity for producing more offspring, not complexity.

In contrast, creative evolution has a built-in propensity for producing new organs of complexity. It solves the problem of the biological arrow of time: evolution proceeds in the direction of making more and more sophisticated expressions of more and more biological functions.

We can still ask: What is the ultimate objective of evolution? Where is evolution going? Or an even more basic question: If evolution is God's creativity, what is God's purpose in evolution? Why create more sophisticated organisms? What is the meaning of this wonder-filled evolving biological universe?

As you know, Christian theologians are usually antievolution. But one glaring exception to this general rule is a Jesuit priest of the 20th century, Pierre Teilhard de Chardin (1961). He not only supported evolution, but he also saw clearly that evolution rises against the march of entropy, to create increasing complexity and order by first creating the *biosphere* and then creating the *noosphere*—the sphere of the evolving mind. Then he proposed that the future of evolution lies in the *Omega point*: a time when godliness becomes dominant. It is easy to see the parallel to the idea of the Second Coming here.

One common thing in the history of creative ideas is that often a truly creative idea nonlocally expresses through more than one visionary. A second visionary to intuit this way even before Teilhard de Chardin was the Hindu mystic-philosopher Sri Aurobindo (1996) in the first half of the 20th century.

Hinduism is quite different from Christianity in its perspective on evolution. The Hindu *puranas* (texts narrating the history of the universe) in the mythology of the *avataras*—descent of God in biological form—can already be seen as depicting evolution. According to the *puranas*, God's first *avatara* is in the form of a great fish. The second is a great tortoise. The third is a boar. The fourth is a man-lion. The fifth is a dwarf man. The sixth through the ninth *avataras* depict an evolution of the human being, from the primitive highly emotional mind to Buddha, a man of mental maturity and emotional equanimity. The tenth *avatara* is yet to come, again alluding to something like the Christian Second Coming (except that for Hindus it is the Tenth Coming).

Of course, on the other side of Hinduism is a general dismissal of the manifest world as illusory and ephemeral, not worthy of one's creative attention, evolving or not. Only the realization of the permanent, of the unchanging reality underneath the manifestation of consciousness as the ground of being, is the highest goal human life can be dedicated to achieving.

So Aurobindo's philosophy was developed with this background. However, what is novel is that Aurobindo integrated the two forces of

Indian thinking with the idea of first *involution* and then *evolution* of consciousness. Ken Wilber (1981) has put further flesh on the skeleton of Aurobindo's work, and so have I (Goswami, 2001). Figure 9-2 shows the evolved version.

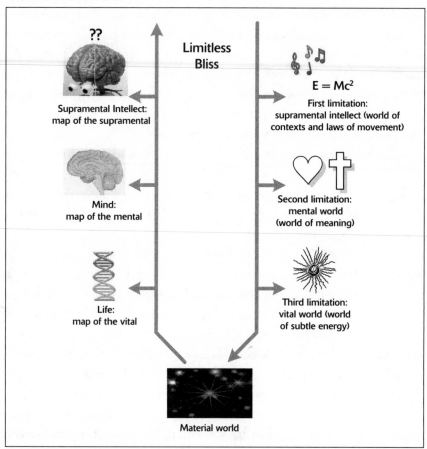

FIGURE 9-2. Involution and evolution of consciousness.

Why involution? Aurobindo anticipated the necessity to see evolution in terms of a science within consciousness, beginning with consciousness as nondual, the ground of being. All possibilities are there—past, present, and future. So there is no time: it is a truly eternal unchanging reality—nothing happens. To make something happen, there have to be limitations. Hence, involution is the imposition of a progressive series of limitations.

128

The game is play, purposive play, a play of expression, expression of all that is possible to express, "to make the unconscious conscious." When you play a game, the first thing you do is to make a set of rules. A game without rules is chaos. God makes man in His own image. As above, so below.

The first involution is to create a limitation of rules, contexts, and archetypes for all the movements and changes to come. This includes the rules of quantum physics; from here on out all permitted possibilities are quantum possibilities. So we now have the *supramental* world of quantum possibilities.

The next stage of involution is the further limitation of meaningfulness. Of all the quantum possibilities, let's restrict to those that are meaningful. This gives us the *mental* world of meaning.

The next level of involution created the possibilities of the *vital* world, the set of morphogenetic fields that help create the particular biological forms that get to play. The subtle cannot collapse of itself, because collapse requires tangled-hierarchical quantum measurement devices that come about only when micro makes up macro.

The final limitation of the involution is therefore the physical, which is made in a special design of micro and macro to first help collapse the quantum possibilities into actualities and then to make software representation of what has gone before: the subtle vital, mental, and supramental worlds.

Evolution begins with the creation of the first living cell. It goes through various stages, more or less, with directionality of more complexity and more order, with the purpose of making more and more sophisticated representations of more and more biological functions, whose blueprints are the morphogenetic fields. This is the evolution of life.

Eventually, the brain's neocortex evolves in biological beings and now the mind can be mapped. Evolution becomes the evolution of the representations of mental meaning. The evolutionary story here is told in the scientific research of anthropology, sociology, and psychology. And undoubtedly, there are signs of evolution, actual stages, in all these disciplines.

So what is the future of evolution in this picture? You can see it very clearly now. Mental evolution culminates with what Jung called *individuation*: when human beings learn en masse to mentally represent and integrate in their behavior all the supramental archetypes. This includes the integration of feelings and meanings, paying attention to the archetypal contexts of both. Aurobindo called this step the *Overmind*. The next step is unimaginably glorious. It consists of developing the capacity to represent in our bodies all the possibilities of the supramental archetypes of mental thinking: love, beauty, justice, good, and all that which we call godliness. Aurobindo poetically called this stage the *Supermind*, bringing down the divine to the earth level.

NOW WHAT?

How do we go from where we are now to there, to explore the supramental in manifestation? We can only speculate. I have indulged in such speculation elsewhere (Goswami, 2001, 2004). In this book I want to take a different track; I want to examine what we can do right now to facilitate the course of evolution. In short, I am proposing *quantum activism*. You have already glimpsed the idea in Part One. The idea is further developed in Part Five.

The evolution of mental representation making is stalled. Evidence is everywhere: in politics, in economics, in business, in education, and in religion, the furthering of meaning processing is no longer taking place. There is perhaps more than one reason for this temporary setback. But two reasons stand out. One is that we have not integrated feelings into our processing of meaning. The other reason is the current materialist paradigm, which is a killer of meaning.

So it is imperative for all thinking people to examine the theory and data presented here for rediscovering God and spirituality in our paradigms and in our lives. I hope it is convincing already. But there is much more to come when we discuss the evidence for the subtle bodies and for more quantum signatures in that context. This is the subject of Parts Three and Four.

One more comment in closing. The evolutionary outlook of God's creation solves one problem that many people worry about—why a good God allows evil to happen. Evolution begins with quite primitive and imperfect representations of godliness, but it gets better. So initially what takes place in the creation appears imperfect, even evil, but in time, as the representations get better, good more and more prevails over evil. A prejudice-free look at the evolution of humanity through history will readily show you the truth of this statement. Over time, we have gotten less violent and more loving. Sure, we have to go further, but the existence of today's imperfections should not dishearten anyone from accepting God.

Part Three

The Evidence for the Subtle Bodies

I n 1994, I was dreaming one night. I heard something. It was like a voice that was speaking to me. The voice grew louder and louder. Soon it became an admonition. I could clearly hear, "The Tibetan Book of the Dead is correct; it's your job to prove it." The admonition was so loud it woke me up.

Ah, so, I took the dream quite seriously. But the task proved very difficult. The Tibetan Book of the Dead—guide to the experiences of the consciousness between death and rebirth—was about survival after death. What are said to survive are "subtle" bodies, but what are they?

The Upanishads (Hindu scriptures) and the Kabbalah told me that the subtle body consists of the vital energy body, the mind, and a supramental body of archetypal themes. But if these bodies were nonmaterial, as The Tibetan Book of the Dead implied, nobody knew how they could interact with the gross physical body.

One thing heartened me a lot. By then I had avidly read scientific research that pointed out deficiencies of the materialist approach to science, and I came across the work of John Searle and Roger Penrose proving to us that computers cannot process meaning; there was scope for nonmaterial mind after all. I was also aware of Rupert Sheldrake's work on morpho-

genetic fields; it became clear to me that the ancient vital body is nothing but the reservoir of morphogenetic fields.

So it was clear that a nonmaterial mind processes meaning and a non-material vital body whose movements we feel guides the creation of biological form. But how did these bodies interact with matter without that dreaded word, dualism?

One day, I was talking with a graduate student whose boyfriend had died. I was trying to say to her by way of consolation that maybe her boyfriend's subtle body—mind, vital, and all that essence—survived his death. Maybe death was not as final as we currently think under the mesmerism of materialist science. Suddenly a thought came to me—suppose the essence of mind and vital body consists of possibilities, quantum possibilities. Could that not solve the problem of dualism as well as that of survival? I was elated.

Chapter 10

The Interiority of the Psyche

Looking at the material world with all the scientific explanations and manipulations that go on today makes you think that the God hypothesis is not needed. Or at best you tend to conclude that God must be behaving as a benign caretaker of the garden He (She) created (a philosophy called *deism*). But as soon as you look inside your psyche, it is much easier to believe in God, materialist behavioral/cognitive psychology notwithstanding.

For one thing, in the psyche, we experience feelings that no one has ever succeeded in explaining with a mechanical model like a feeling computer; we don't even know how to begin. The other aspects of our "inside" experience of the psyche are thinking and something we call "intuition," although there is much misunderstanding about it. Is the God hypothesis needed for understanding our feeling, thinking, and intuitive faculties? Are feeling, thinking, and intuition really impossible problems for the materialist approach?

There are now computer programs available, in the category of "artificial life," that can perform some of the well-known functions of life—self-maintenance and self-reproduction, maybe even some rudimentary evolution. What test of life can you give these programmed lives that they will fail?

When you go to a restaurant and are taken aback by the beauty of an indoor plant, how do you determine whether it is real or artificial? You may try to decide by the feel of the plant, by touching it. If you are sensitive to vital energy, then you don't even need to touch it. If the plant is alive, you can tell by its "feel," even at a distance. How is this possible? Through the nonlocal consciousness of the vital energies of the live plant, of course.

No artificial life researcher can ever claim that artificial life mechanisms can give you such a feel. Nor can a materialist biologist ever explain the origin and nature of feelings that you experience, except to mumble something about the possibility that your feelings may have evolved because there is some evolutionarily adaptive value to them.

There are also the artificial intelligence researchers who claim that thinking is nothing but computing and that computer programs can be built to simulate thinking. Initially there seems to be some substance in this claim, but ultimately it is deceptive. There is an essential aspect of thought—i.e., meaning—that materialists have hitherto missed. Meaning is an impossible problem for materialists and, as we will see later, it requires the God hypothesis for it to be included in the equation of our experiential being.

Intuition, as I said above, is sometimes misunderstood, but when we remove the chaff from the wheat, what remains is intuition's ability to connect with the contexts of feeling and thinking that Plato called *archetypes*, such as love. Our intuition of the archetypal contexts of feeling and meaning gives value to these latter experiences.

Materialist models for the archetypal experiences are terribly inept. For example, there is a book called *The Selfish Gene* (Dawkins, 1976) that tries to establish a biological theory that altruism for another person depends on the amount of genetic material you share with him or her. This theory would not be so lame if the author didn't claim that this "concern" is all there is to altruism. Need I say more? Sometimes materialists talk like computer programs.

On the other hand, feelings that we intuit, such as love, have always been recognized as divine qualities. Are they? This is the subject of a later chapter.

What is striking about our experiences of vital feeling, mental meaning, and the archetypal contexts of feeling and meaning is that sometimes they occur with such depth and immediacy that all doubt disappears. In those moments we know not only that there is God, but also that we are It. I am talking, of course, about mystical experiences that prompt mystics to tell us to look for God within. "The kingdom of God is within you," said Jesus. Mystics claim that these experiences are also of our own true self; when we are situated in this inner self, we can directly feel that we are the children of God.

When we look at our experiences of feeling, meaning, and the archetypal contexts of feeling and meaning through the conceptual lens of the new science—science within consciousness—we find that there is ample experimental proof that they don't arise from the physical body. They occur in conjunction with the body, but they are not of the physical body. Instead they come from God, or more accurately from the Godhead; we choose them from our own God potentia. In other words, no mystic has to tell us that God is our "father." Every one of us has that intuition already. The new science is just validating that intuition.

One more impossible problem for the materialist paradigm is its utter inability to distinguish between inner and outer awareness. That paradigm of reality is entirely based on the material world that we experience outside us. In that worldview, every interior experience is meaningless epiphenomena of matter, of the brain. It must be, or else it causes a paradox. So materialists denigrate inner experiences as subjective, untrustworthy, and of no causal consequence—although they acknowledge that there must be some adaptive value to them so that they can evolve through natural selection.

Can putting God back into science enable us to understand why some of our experiences are outer while others are inner? Yes. Psychology, the science of our psyche, our inner experiences, has become marginalized by materialist beliefs that permit only the narrow cognitive/behavioral domain of psychology. Literally, a scientific revision of God is needed to reclaim an interior psychology and bring it back to full academic standing.

THE DISTINCTION OF INNER AND OUTER

Finally, we come to the quintessential problem of this chapter—the distinction of inner and outer experience of our awareness. The materialists have no possible explanation for the inner experience, so they just wish it away as subjective epiphenomena needing no explanation. But idealist philosophers, who do value the inner experience, don't do very well on this question either. They just make the inner nature of the psyche a matter of metaphysical truth: it's the way psyche is. But in idealist philosophy, consciousness is the ground of being; all things are inside consciousness—matter and psyche. So, then, why do we experience one as outside and the other as inside?

The quantum nature of the stuff of the psyche—the mind, the vital body, and the supramental—is why these experiences are inner experiences. Quantum objects are waves of possibility, expanding in potentia whenever we are not collapsing them. When we collapse a mental meaning wave, a particular meaning is chosen and a thought is born. But as soon as I stop thinking, the wave of possibility goes expanding again. So between my thought and your thought, the wave of meaning expands so much and becomes so many possibilities that it is highly improbable that you will collapse the same thought as I. (An exception occurs when we are correlated, as in mental telepathy. Another exception sometimes occurs when two people of similar conditioning converse.) So, generally speaking, thoughts are experienced as private and therefore as inner.

Now compare the situation with material objects. There is a fundamental difference between the subtle bodies and the gross material body, which is why such names are given. The subtle bodies—the vital, the mental, and the supramental—are all one thing; they are indivisible. But, as Descartes recognized, matter is *res extensa*: extended body. Matter can be subdivided. In the material realm, micro matter makes up conglomerates of macro matter.

So although quantum physics rules both domains of matter, micro and macro, there is a spectacular difference that arises when we consider macro matter as massive conglomerates of the micro. A massive macro body's wave of possibility becomes very sluggish.

Suppose your friend and you are looking at a chair. If you collapse the chair's possibility wave in order to look at it, fine. You have done it and you see the chair over there by the window. Soon after, your friend also looks. In between your collapse and your friend's collapse, the chair's possibility wave expands no doubt, but very little. As a result, when your friend collapses it, the new position of the chair is only different by a minuscule amount from where you observed it, imperceptible without the help of a laser instrument. As a result, you both think you are looking at the chair in the same place; you have a shared experience, so the chair must be outside of both of you.

The macro-material world is built in this way. And this is useful, because otherwise we could not use it as a reference point. If your physical body were always depicting the uncertainties of quantum movement, who would you be?

Also, if the quantum nature of macro matter were not subdued, how could we use matter for making representation? Imagine writing your thoughts on a white board with a magic marker, only to see the marks move away in subsequent collapsed events. What would that do to our representation-making capacity?

Finally, matter needs micro-macro division in order to have tangled-hierarchical quantum measurement in which the process of amplification from micro to macro is tangled, containing an irreducible complexity, a discontinuity. (See chapters 6-8.)

So, both matter and mind are inside us, but the micro-macro division of the material world camouflages the quantum nature of matter. The macro world of matter—all that we can see directly—behaves almost like a Newtonian object, giving us the illusion of a shared reality, the illusion of objectivity. Materialist science grew out of this illusion, but that is the downside. The upside is more important. We can use matter to make representations of the psyche.

Modern psychology began with the study of the psyche; initially, the inner experience and introspection were what constituted psychology. Understanding the inner experiences, such as conscious awareness and the self that seems to organize the conscious experience, was the goal. Also, from the start, one of psychology's major applications was clinical, to help people with their mental health. Only in the 20th century did John B. Watson and B.F. Skinner, under the aegis of materialist philosophy, begin to undermine the inner experience in favor of behavior studies. Brain was the black box of these early studies, and the psyche and inner experiences were relegated to secondary epiphenomena. Later neurophysiology, cognitive science, and artificial intelligence research were added to the behavioral studies, making the subject more interesting. Unfortunately, the reshaping of psychology in this manner took this science further and further away from its avowed goal of understanding the inner experience, which is essential for alleviating people's suffering.

Our analysis with the new science is showing that the behavioral-cognitive science and neurophysiology, by pursuing the study of the representation-making apparatus and the representations, is following a goal complementary to the original goal of psychology. These studies are important, no doubt, but of limited use when the representations go awry, as in mental illness. The representations also are of limited use when we need to make new creative representations driven either by environmental pressure or by the creative urges of the divine wanting to express Itself more fully.

Fortunately, there are other forces in psychology that continued to pursue the original goal. Two such strains can be identified: the depth psychology that began with Freud's discovery of the unconscious, and the humanistic/transpersonal thread that began with the human potential movement. In subsequent chapters, we will see that these branches of psychology are positing new challenges, new impossible problems for the materialist view of reality. And when we solve these problems by looking through the quantum lens, we find more evidence for God.

IS INNER EXPERIENCE SECONDARY TO OUTER EXPERIENCE?

Materialists say that inner experience is not important, that it is secondary to outer experience. For one thing, it revolves around our outer experience. This is a point well taken. Indeed, if you monitor your thoughts and feelings for a day, you will be surprised to find how much of the inner experiencing is just a regurgitation of the outer world or a reaction to it.

But don't get fooled by the claim that inner experience is secondary to outer experience. That inner reaction starts from your habit of ascribing meaning to the outer experience and taking it too seriously. But that is not all there is to the inner experience.

Let's consider dreams. We live a substantial part of our nightlife in the dream world. This is the closest we come to living in an internal world. If inner experiences are important, then that importance must show up in dreams.

Here also, some have made the case that the outer experience dominates even our dreams. Many of our dreams are called *day-residue dreams*, because they are nothing but a review of what went on during our waking hours. Add to this what every philosopher knows as a fatal flaw of dreams: they seemingly have no continuity. How can dreams, our dream lives, be significant if they lack continuity?

Again, think in terms of meaning. Dreams are about meaning; the dream life is to be distinguished from our waking life in that its significance and continuity arise from processing meaning. If you keep track of the meaning as it unravels in your dreams for a while, you can easily prove to yourself that there is continuity. Dreams, in fact, are an ongoing commentary on how our lives are playing out at the meaning level. You may have to penetrate the rich symbology of your dreams to analyze their meaning, but it's worth it. You will find that there is much more to dreams than the residue of the outside world. There are vital, mental, and supramental dreams as well; all the realms influence our inner experience, not just the physical.

HOW INNER EXPERIENCE PROVES THE EXISTENCE OF GOD

To summarize, please note the logical steps through which the existence of God is established from our inner experience:

Without our vital body giving us feeling, our mind giving us meaning, and our supramental body giving us values, there would be no feeling, no meaning, and no value to anything, even for scientific work or even for materialism. Since feeling, meaning, and value are essential aspects of our inner experience, the importance of these bodies—vital, mental, and supramental—cannot be refuted.

If we say that feeling, meaning, and value evolve from matter because of our survival necessity, then these qualities of our experiences would be ornamental epiphenomena of matter. But they are not, by two criteria. First, matter cannot even process them. Second, we see the evidence for the causal efficacy of the perception of feeling, meaning, and value in creativity and spirituality, and in dreams, in disease and healing, in love, and in events of synchronicity (chapters 11-19). So feeling, meaning, and value are not the products of Darwin's black box—evolutionary adaptation.

We have to recognize that the inner life can be the focus of living, equal to our culturally chosen outer life. Australian aborigines and mystics all over the world prove this point empirically.

If our inner experience is causally efficacious and as potent as our outer life, then we must find a scientific explanation for it. Otherwise, science just loses its relevance.

No materialist explanation can be given for the distinction between outer and inner in our experience.

A dualistic explanation—inner and outer as separate realities—is untenable because of experimental data that establish the law of conservation of energy.

A nondual explanation of the inner-outer split of awareness can be given, if we assume that both experiences originate from the collapse (by quantum consciousness or God) of quantum possibilities in consciousness (or Godhead). A spectacular empirical piece of evidence for this is the phenomenon of synchronicity, which anyone can verify (chapter 12).

Proof complete!

AN END TO THE CARTESIAN SPLIT?

Ever since René Descartes recast reality as an internal/external dualism of mind and matter, Western philosophy has been saddled with this split. Even great thinkers such as Immanuel Kant and Ken Wilber, seem unable to jump out of this philosophical box.

Wilber has a lot of influence today on the future of consciousness studies, so let's examine his work in some detail.

Wilber began his career as a philosopher. He endorsed the perennial philosophy (which is another name for monistic idealism) and he very capably translated and clarified its message toward developing a transpersonal psychology for our times. So impressive was he in his earlier expositions that some declare him the Einstein of modern psychology.

And yet, when Wilber focused the direction of his research on developing an integral psychology, he took the Cartesian interiority/exteriority dichotomy as his starting point. The materialist approach to psychology—made up of cognitive psychology, behaviorism, and neurophysiology—is objective, a study of consciousness as third person, "it" and "its." His earlier transpersonal approach, based on perennial philosophy, was directed toward finding the nature of self (I) and, as such, it was a study of consciousness primarily in the first person and secondarily in the second person (I/you and we) when nonlocality of consciousness is recognized. The objective study of "it" and "its" is done in our exterior consciousness, whereas consciousness in the first and second persons can be studied only from the interior vantage point. Hence Wilber's famous four-quadrant model of consciousness studies (figure 3-1, page 45).

But there is no integration as of yet. There is mind and body in the study of consciousness from the vantage point of interiority, but body is now relegated to being an epiphenomenon of the mind. Likewise, there is mind and body from the vantage point of exteriority as well, but mind is looked upon as an epiphenomenon of the body. It does not seem that either vantage point can ever do equal justice to both mind and body.

What, then, is Wilber's solution? Wilber (2000) says that in order to resolve the mind-body dualism, we have to develop our conscious-ness to grow the capacity to experience nonordinary states: "you must further develop your own consciousness if you want to know its full dimensions." Only from the nonrational vantage point of nonordinary "higher" states of consciousness is the mind-body dualism resolvable. Wilber flatly declares that there is no rational solution to the mind-body problem: "this solution . . . is not satisfactory to the rationalist (whether dualist or physicalist)."

I mention Wilber's theory only to make the point that it is extraor-dinary that the quantum/monistic idealism approach does give a rational resolution of the mind-body problem and the interiority/exteri-ority dichotomy that perpetuates it. Quantum physics allows us to see that, like the Newtonian fixity of the macrophysical reality and the behavioral nature of the conditioned ego, the interiority/exteriority dichotomy is also nothing but a camouflage. As we penetrate the cam-ouflage, we extend science to our subjective, interior experiences. It is about time.

Chapter 11

The Evidence for the Vital Body of God

Do we really have a vital body, a reservoir of blueprints of biological form-making? In other words, is there any independent evidence for the vital body, apart from its function in biological creativity as shown in chapters 8 and 9? The answer is yes. The morphogenetic fields give us a profound explanation of feeling: what we feel, how we feel, and where we feel. To be sure, this is experiential, but a second and more objective piece of evidence for the vital body arises from its importance in alternative medicine. A third is the highly documented and very practical application of *dowsing*. We discuss all this in this chapter.

Sometimes, it is hypothesized that feeling and emotions are the territory of the neurochemistry of the limbic brain. To this end, the researcher Candace Pert's (Pert, 1997) experiments on the "molecules of emotion" (one example is the endorphins) are important. Certainly those molecules are telling us something, but it should be obvious that molecules are material correlates of feelings rather than their cause. Just because two things occur together does not guarantee that one is the cause of the other.

Fortunately, in the psychology of the East, feelings are recognized to be associated with the physiological organs, and emotions are clearly seen as effects of feelings on the mind and the body physiology. According to Eastern metaphysics, there are seven major energy centers of the body—the *chakras*—where we feel our feelings. But through the centuries, although the idea of the chakras has found much empirical validation from spiritual disciplines, not much theoretical understanding has come together. Now, finally, with the idea of Sheldrake's morphogenetic field (the source of the programs that activate the genes of the cells belonging to different organs, resulting in cell differentiation), an explanation can be given for the chakras, where feelings originate.

This subject then becomes evidence for the existence of God, because without the God hypothesis and downward causation, we cannot incorporate the morphogenetic fields in science without implicit dualism.

MORPHOGENETIC FIELDS AND THE CHAKRAS

I have examined this subject in some detail elsewhere (Goswami, 2004), so I will be succinct here. You can discover for yourself how a little quantum thinking enables us to scientifically theorize. First, look at the major chakras (figure 11-1) and notice that each of them is located near major body organs of biological functioning. Second, make a note of the feeling you experience as you concentrate on each of these chakras; feel free to use your memory of past feelings. Third, realize that feelings are your experiences of the chakra's vital energy—the movements of your morphogenetic fields, which are correlated with the organs of which they are the blueprint/source. This leads you to the inevitable conclusion: chakras are points on the physical body where consciousness simultaneously collapses the movements of important morphogenetic fields along with the organs of the body that represent them. Now was that so hard?

It may be of interest to know the literal meaning of the Sanskrit word "chakra." It means wheel, circularity—implicitly a reminder that a

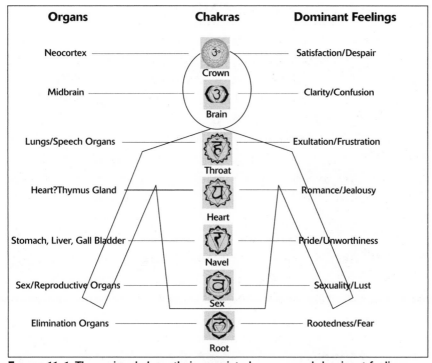

FIGURE 11-1. The major chakras, their associated organs, and dominant feelings.

tangled-hierarchical quantum collapse ensures the arousal of self-reference at each of the chakra points. Our new science is validating ancient intuitional wisdom.

The materialists have it all reversed. They think that we feel emotions in the brain—that emotions are brain epiphenomena, and then the emotions come to the body through the nervous system and the so-called "molecules of emotion." But actually, it is the other way around. We feel feelings at the chakras first, then the control goes to the midbrain for integration through the nervous systems and the "molecules of emotion," and eventually the neocortex gets involved when mind gives meaning to the feelings.

But where is objective data on all this? Chakras are fun to experience, you may say, but is there any solid experimental data proving their importance and hence the importance of the movement of vital energy? There is indeed. This is the subject of chakra medicine.

CHAKRA MEDICINE

The fundamental idea of chakra medicine is that for good health, the vital energy should be experienced in a balanced manner—going in and coming out, and not producing any ongoing excesses or deficits in any of the chakras. If there is an imbalance or a block (suppression) in the movement of vital energy at any chakra, the corresponding organ or organs malfunction, and eventually we get disease. Healing a chakra consists of restoring balance to the energy flow at the chakra. There is much data now to confirm this, thanks to physicians like Christine Page (1992) who have been engaged in chakra medicine and have accumulated quite a few success stories.

It is also interesting that the chakra imbalances are often produced because the mind intervenes. For example, the general mental suppression of heart chakra energy, especially for the American male population, may be responsible for the malfunction of the immune system, causing cancer. Since mental processing is involved, we will discuss this research further in later chapters. (See also Goswami, 2004.)

VITAL BODY MEDICINE: TWO SPECTACULAR
EXAMPLES OF IMPOSSIBLE PROBLEMS FOR THE MATERIALIST

When President Richard Nixon went to China in 1971, the visit renewed not only trade with that country but also an interest in traditional Chinese medicine, especially acupuncture—healing through the superficial insertion of sharp needles into various parts of the body called *acupuncture points*. How acupuncture heals has become a major problem for allopathic medicine, based entirely on the materialist approach to reality.

Aallopathic medicine researchers look for an explanation of acupuncture along materialist lines of thinking. For example, one theory of how acupuncture relieves pain is that the inserted needles cause minor tolerable pain to take attention away from the major source of pain by jamming the communication channels of the nervous system.

It turns out, however, that traditional Chinese medicine (Liu, Vian, and Eckman, 1988) already had the right explanation: the movement of a mysterious energy called *chi* as the vital energy movements of the vital body. We must, of course, allow the allopathic allergy (called dualism) toward the vital body and chi to heal through the quantum medicine of *psychophysical parallelism* (figure 1-6, page 26).

In the Chinese view, chi has two modalities: *yang* and *yin*. You can see the parallel with quantum thinking here. Yang is the analog of the wave mode of chi, and yin is the analog of the particle mode. Correspondingly, each organ, and even the entire organism, is classified according to the modes in which chi is processed: the yang mode, which involves creativity, and the yin mode, which involves conditioning.

We now can understand how acupuncture relieves pain. For a body with healthy organs, the application of the needles to suitable areas stimulates the correlated parts of the vital body. This stimulation produces the flow of creative yang chi through vital pathways called *meridians* to the vital blueprints of the major organs. This increases the general level of yang chi in the vital body, especially in the vital correlates of the brain areas that produce endorphins, the brain's own opiate pain reliever. In other words, the manifestation of the vitality of chi at the vital level manifests brain states with endorphins.

One can verify the correctness of this picture by injecting endorphin blockers, antagonist narcotic drugs that block the action of opiates. Indeed, this neutralizes the pain-relieving effect of an acupuncture treatment. (For more details, read Goswami, 2004.)

If acupuncture is a relatively recent thorn in the side of materialist allopathic medicine, homeopathy is an old thorn. The major mystery of homeopathy is its "less is more" philosophy. In homeopathy, the medicinal substance is diluted with a water-alcohol mixture to such infinitesimal proportions that, on the average, not even one molecule of the medicinal substance can be said to be present in the concoction that is administered to the patient. If no medicine is given, how does homeopathy heal?

Allopathic medicine attacked the problem by running many clinical tests to directly disprove the efficacy of homeopathy. But tests have now confirmed that homeopathy does work, and not as a placebo either—the effect is not based only on the power of suggestion and belief.

How does homeopathy work? First, we must recognize that the medicinal substances in homeopathy are organic, having not only physical bodies but also vital bodies. The physical body part is diluted away (which is good, because usually this part is poison to the human body), but the vital body is preserved.

How is the vital body preserved in the eventual medicine? To solve this mystery, we need to look at the details of how a homeopathic medicine is prepared (Vithoulkas, 1980). You take one part of medicinal substance and dilute it with nine parts of the water-alcohol mixture. Then you take one part of this dilution and dilute it again with nine parts of water-alcohol mixture. You do that 30 times, one hundred times, and even a thousand times to get homeopathic medicine of increasing potency.

The procedure sounds innocuous—until we realize that we have missed something important. At each stage of dilution, the mixture is thoroughly shaken. The word *succuss* is used for all that shaking—and herein is the mystery. *Succussion* is the process of forcefully striking a homeopathic remedy against a firm surface. The succussion transfers, through the intention of the preparer, the vital energies of the medicinal substance to the water of the water-alcohol mixture, which becomes correlated with the vital energy of the medicinal substance. As a result, whenever we take the homeopathic medicine, although we don't get any of the physical part of the medicine, we do receive the vital part with the water.

So if the disease is at the vital level, due to vital energy imbalances, then homeopathy will work better than allopathic medicine. This is because it addresses the vital energy imbalance directly through the application of the vital energy of the remedial medicine, which is chosen by the second principle of homeopathy, "like cures like." If the medicinal substance incites the same symptoms in a healthy body as the symptoms of the unhealthy body, it must mean that the vital energy movements of the medicinal substance and the body's relevant (unbalanced) vital energy movements are in "resonance." In that case, the vital energy of the medicinal substance will balance the imbalance of the unhealthy person's vital energy.

These main principles of homeopathy and many more details can be understood using vital body science. (Read Goswami, 2004.)

THE VITAL BODY AND THE EXPLANATION OF DOWSING

Dowsing is a very well-known practical phenomenon that many people have used to locate underground water. It comes in handy when you are digging a well and you don't want to rip up your entire backyard. Dowsers are talented people who let a *divining rod* in their hands guide them to the source of water.

Materialists have so many seemingly insurmountable problems with dowsing that they feel compelled to denounce it as pure chance. The dowser is a lucky con artist of some sort, they may claim. But if dowsers are operating by mere chance, it's pretty amazing how they can be so successful.

With an understanding of vital energy phenomena and how vital energy works, one can formulate a theory to explain dowsing. Conversely, our theory construction enables us to claim that dowsing is a very convincing proof of the potency of vital energy. Dowsing also may direct us to new ways of harnessing downward causation and other fruits of the new science thinking.

Of crucial importance in dowsing is the dowser's intention to find water. The intention acts through the feelings of vital energy that connect to the vital energy correlated with the underground water.

For me personally, the idea that water is capable of holding vital energy correlations in memory was the stumbling block that kept me from understanding dowsing. But once I recognized how homeopathy works through the process of succussion, understanding dawned. Water has a long history with us, and underground water can easily be succussed with vital energy in many ways; it is not important to know how. If the dowser's intention is good, the vital energy of the underground water is going to become correlated with the vital energy of the dowser. This is the first step.

There is still a crucial piece missing. Why does the dowser use a divining rod? Wouldn't it make more sense to use his or her hands?

Perhaps the divining rod just helps to focus the dowser's intention on the target and gives it directionality?

The mystery is clarified by the important research on intention done by the engineer-author William Tiller and his collaborators (Tiller, Dibble, and Kohane, 2001). After many years of experiments, Tiller has demonstrated quite convincingly that the causal potency of human intentions can be transferred to material objects.

So the dowser transfers the intention of finding the water to the divining rod, which then becomes the instrument of the intention and facilitates carrying it out. Phenomenon understood and explained.

Here is a practical application of dowsing that you may try. If you frequently visit health food stores, you must be aware of the many choices among all the different herbal "energy" items with their extravagant claims. Is there any way to choose what you need intelligently? Using dowsing, you may quite easily find the one that has a good amount of vital energy. For a dowsing rod, you can make your own contraption, using two metal rods and loosely attaching them at a fulcrum so that the rods can freely rotate about the fulcrum. Now hold your divining rod loosely. For a starter, verify that if you approach a nonliving object with your dowsing rod, nothing happens; the two rods do not separate. Now aim the rod at the herbal concoctions one at a time and repeat your experiment. You should find that with some of them, the two rods will separate without any effort on your part. Obviously, you should choose from among those. Note: do not forget the importance of intention in your trials.

One last passing comment. You see how important Tiller's research is for future technology based on downward causation. It is very important for other researchers to replicate this research.

To Recap: What Is the Evidence for God in All This?

Impossible problems require impossible solutions. We began this chapter with a discussion of morphogenetic fields, the blueprints that our designer of reality, God, uses for making the designs of life. Morphogenesis is an impossible problem for the materialist because of the nonlocality involved in its processing. And when we are able to connect the movement of morphogenetic fields and the organs they help develop with our feelings, we have found the explanation of still another impossible problem for materialists—feeling.

So what we garner from this chapter is that the God hypothesis is needed to incorporate feelings as part of our experience. You will notice that feeling-oriented cultures tend to be believers in God (good or bad), whereas when rationalism dominates a culture, it tends to move away from the God hypothesis. This is not a coincidence.

Materialist science is practiced with gusto today because it seems to give us control over our environment. But feelings are something that we cannot control. If we try to do so, it is at our own peril; witness all the diseases that we develop when we suppress feelings.

Chapter 12

Exploring the Mind of God

s *mind* the mind of the brain, or is it the mind of God (or Godhead)? We can now deal with this question in some detail.

The neocortical part of the brain, supposedly the site of the mind, is a computer of sorts. So materialists ask, "Can we build a computer with a mind?" If we can, that would prove that the mind belongs to the brain.

Can a computer simulate mental intelligence? This question originated an entire field of study called *artificial intelligence.* The mathematician Alan Turing formulated a theorem purporting that, if a computing machine can simulate a conversation intelligent enough to fool someone into thinking that he is talking to another human being, then we cannot deny a computer's mental intelligence.

In the 1980s, there was a telephone number in Canada that you could call to talk to a computer simulation of a California psychiatrist. Many people talked to the computer and later admitted that they could have been fooled, so authentic was the machine with the touchy-feely psychobabble of the day.

So have computers passed the Turing test? A computer has beaten one of the world's greatest chess players in a game of chess, so maybe the computer is even more intelligent than the human being. After all, not only have we built a computer with a mind, we have built a computer with a mind better than one of our best.

"Not so fast," said a philosopher named John Searle. In the 1980s, Searle constructed a puzzle called the Chinese Room to make an argument against the so-called intelligent computer.

Imagine yourself in a room, no doubt wondering what to do, when a card comes out from a slot. You take the card and see scribbles written on it that look like Chinese to you. But you don't know Chinese, so you don't understand the meaning of what is written on the card. Looking around, you see a sign in English telling you to consult an English dictionary, where an instruction is given for finding a response card from a pile of cards. You carry out the instructions, find your response card, and put the card in an outgoing slot as instructed.

So far, so good? But now Searle will ask you this: "Do you understand the purpose of this trip into the Chinese room?" When you admit to being a little puzzled, Searle explains, "Look, you could process the symbols inside the room just as a computer does. But did that help you process the meaning of what was written?"

So this was Searle's point. A computer is a symbol-processing machine; it cannot process meaning. If you think we can just reserve some symbols to denote meaning, think again. You will need more symbols to tell you the meaning of the meaning symbols that you have created. And so on ad infinitum. You need an infinite number of symbols and machines to process meaning. Impossible!

Searle wrote a book, *The Rediscovery of the Mind,* in which he suggested that a mind is needed to process meaning; the brain alone cannot process meaning, but can only make a representation of mental meaning.

Later the physicist/mathematician Roger Penrose gave a mathematical proof that computers cannot process meaning. The name of his book, *The Emperor's New Mind,* is equally provocative. Like the emperor's new clothes in the famous story, all the hoopla notwithstanding, the computer's new mind is imaginary.

What Searle and Penrose have accomplished is very good science, because their work completely negates the materialist biologist's contention that meaning is an evolutionarily adaptive quality of matter. If matter cannot even process meaning, how can matter ever present any meaning-processing capacity for nature to select, survival benefit or not?

So the mind does not belong to the brain; it is independent of the brain, being what gives meaning to our experiences. But how does it follow that it belongs to God, that it is God's mind?

We have no doubt that the brain and mind work together; memories are stored, right? But if they are totally different, brain being matter substance and mind being meaning substance, how do the two interact? How do they work together? They need a mediator.

So God is needed, a quantum God: God as quantum consciousness. If mind and brain both consist of quantum possibilities of consciousness, mind being meaning-possibility and brain being matter-possibility, then can you see that God can mediate their interaction? God-consciousness collapses the possibility waves of both brain and mind to experience mental meaning, at the same time making a brain memory of it (figure 12-1).

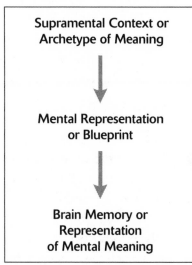

Supramental Context or
Archetype of Meaning

Mental Representation
or Blueprint

Brain Memory or
Representation
of Mental Meaning

FIGURE 12-1. How a supramental context of meaning is represented in the brain through the intermediary of the blueprint of the mind.

You can still argue that this is just theory. Where is an experiment? In experimental science, the prediction of a negative result is often as good as that of a positive one. We have a negative experimental test here: computers cannot process meaning. It is a fact that so far no computer scientist has been able to build a meaning-processing computer to refute our test hypothesis. In the least, this is a prediction of the theory that will never be falsified, I guarantee you.

Is There Anything Practical about Meaning?

There are other phenomena that provide proof that meaning is important to us, that processing meaning properly is good.

When we ascribe wrong mental meaning to our experiences, we feel so separate that it can make us sick (Dossey, 1992). For example, we may feel love in our heart (chakra), but think it is inappropriate to express it or not know how to express it appropriately. As a result of the inadequacy of our perception of meaning, we suppress our feeling. This meaning-induced suppression of feeling at the heart chakra can block the free flow of vital energy there, so much so that the correlated actions of the immune system (through the agency of the thymus gland) may also be blocked. And this has been known to lead to cancer. When we learn to love, giving it proper mental meaning, and are able to express it, the blocks lift and we are healed. This, too, has been documented. (See, for example, Goswami, 2004; see also chapter 18.) So this is one kind of practical data on meaning.

In this way mental meaning is not just theory. There are two other very definitive objectively testable pieces of evidence for the practicality of meaning and meaning processing. These are the phenomena of creativity and love. But these also involve the supramental in a major way, so I discuss them in separate chapters (chapters 16 and 17).

A substantial amount of sleep time is spent in an altered state of consciousness that we call dreaming; this is objectively documented by showing that the brain waves change between wakefulness, deep sleep, and dreaming. Although dreams are usually experienced subjectively, there are objective consequences of dreams that can be measured objectively. There are physical explanations of dreams that have been proposed, but they fall short because they cannot explain why dreams should make any tangible measurable difference in people's lives. I will take up the subject of dreams in chapter 14.

Although thoughts and dreams are ordinarily experienced internally, as private and subjective experiences, there are occasions when two people share thoughts, even dreams. This is the subject of telepathy, which is shared and therefore public, entirely subject to objective testing. There is now substantial data on telepathy, even dream telepathy. (See chapter 16.)

SYNCHRONICITY

Another phenomenon in which meaning plays a central role is synchronicity (Carl Jung's term). I have mentioned synchronicity earlier; it is an acausal but meaningful coincidence of one external event and one internal event. Is this meaning of the coincidence merely subjective, with no experimentally verifiable consequences? Often the perception of the meaning causes observable life changes in the perceiver that can in principle constitute objective evidence.

An example from Carl Jung (1971) will show how special synchronicity experiences can be. Jung was dealing with a client, a young woman, who was "psychologically inaccessible" with "a highly polished Cartesian rationalism with an impeccably 'geometrical' idea of reality" and who did not respond to Jung's repeated attempts to "sweeten her rationalism with a somewhat more human understanding." Jung was desperately hoping that "something unexpected and irrational would turn up" to help him to break through the woman's intellectual shell. And then the following synchronistic event took place:

> I was sitting opposite her one day, with my back to the window.... She had had an impressive dream the night before, in which someone had given her a golden scarab—a costly piece of jewelry. While she was still telling me the dream, I heard something behind me gently tapping on the window. I turned around and saw that it was a fairly large flying insect that was knocking against the window pane.... I opened the window immediately and caught the insect in the air as it flew in. It was a scarabaeid beetle, or common rose-chafer (cetonia aurata), whose gold-green color most nearly resembles that of a golden scarab. I handed the beetle to my patient with the words, "Here is your scarab."

This synchronistic appearance of the "dream scarab" in this

patient's inside awareness and the beetle/scarab in her outer awareness broke through the young woman's intellectual shell and she became psychologically accessible to her therapist, Jung.

Synchronistic events like this often happen to those in need of a breakthrough in connection with romance, therapy, and creativity, just to name a few contexts.

Now notice the most important aspect of synchronicity. The simultaneous occurrence of two coincidental events, one outer and the other inner, yet connected by meaning, could mean only one thing. The source of such events must lie in an agency (Jung called it the *collective unconscious*) that transcends both outer and inner, both the physical and the psyche. In the quantum view, this agency is consciousness or Godhead, of which both matter and psyche are quantum possibilities. You can see that Jung anticipated the quantum resolution of mind-body dualism long ago.

More explicitly, in Jung's thinking synchronistic occurrences can be traced to objects of the collective unconscious that Plato called *archetypes*. Jung realized that the archetypes have a psychoid nature, manifesting both outside in the physical and inside in the psyche. These archetypes of our collective unconscious are the contexts of physical laws and mental and vital movement that we have previously called the supramental. So Jung's collective unconscious is connected with the supramental domain in us.

If you want to incorporate quantum consciousness in your life, synchronicity offers you a viable means. Let me mention some examples of how creative people use synchronistic experiences.

Item: Hui Neng, the sixth patriarch of Chinese Chan Buddhism, was in the marketplace and heard somebody reciting what is known as *The Diamond Sutra*, a Buddhist text with the line "'Let the mind flow freely without fixating on anything." He was immediately enlightened.

Item: Alexander Calder, the pioneer of mobile sculpture, was in Paris and visited the studio of Piet Mondrian, the abstract painter. In a flash he thought of using abstract pieces in his moving sculpture.

Item: When Albert Einstein was ill in bed at age five, his father brought him a magnetic compass. Seeing the needle of the compass pointing to the north, no matter how he turned the case containing the magnet, gave Einstein the sense of wonder that pervaded his scientific work.

Item: The Nobel laureate poet Rabindranath Tagore saw raindrops falling on a leaf. At once two sentences of a little verse rhymed in the original Bengali came to his mind. The verse can be translated thus: "It rains, the leaves tremble." Later Tagore (1931) wrote about this experience as follows:

> The rhythmic picture of tremulous leaves beaten by the rain opened before my mind the world which does not merely carry information, but a harmony with my being. The unmeaning fragments lost their individual isolation and my mind reveled in the unity of a vision. (p. 93)

Chapter 13

Soul Evidence

In a bygone era, many scientists were deeply religious and they talked about God quite openly. Einstein was one of them. He was famous for saying things like, "I want to know the mind of that One (meaning God)" or "I can't believe that God would play dice with the universe." The reason for this God talk is misunderstood today. Some scientists think that it is just a casual manner of speaking that was common in those days. Others flatly declare that scientists of that ilk had not yet shed their superstition. But the actual reason for belief in God for Einstein and other scientists like him goes deeper.

It is a fact that science, especially physics and chemistry, is based on laws. But how did these laws originate? And often these laws are expressed in the language of mathematics. What is the origin of mathematics?

If everything arises from the motion of matter, then the laws of physics and the language of mathematics must follow from the random lawless motion of the elementary particles. To the credit of materialists, some attempts at rectification have been made. Unfortunately, no one has succeeded in deducing any physical laws from the random movement of elementary particles. Nor has there been any breakthrough in understanding the origin of mathematics beginning with randomly moving matter.

So the scientists of Einstein's ilk who were reverent about God were no fools. Being good philosophical thinkers, they figured out that the laws of physics and their mathematical language offer definitive proof for God. To be sure, these scientists also believed in Newtonian determinism. Accordingly, they genuinely believed that God created the laws of the universe (along with the language of mathematics), set the universe in motion, and then let the laws dictate the course. This is the reason that Einstein said, "The most beautiful and profound emotion we can experience is the sensation of the mystical. It is the power of all true science."

To these scientists, God was a benign caretaker of the world, a parent who refrained from interfering. To be sure, Einstein never understood the entire message of quantum physics, although he contributed crucially important ideas to it. His comment, "I cannot believe that God plays dice with the universe," came later from his utter frustration with the majority of scientists following the so-called statistical interpretation of quantum physics. Physicists hypnotized themselves, calling the waves of quantum objects "probability waves" and not what they really are, "waves of possibility." Thinking of quantum objects as waves of possibility sooner or later will raise the question in your mind, "Whose possibility?" Instead, physicists ignored the observer effect and remained satisfied with calculating probabilities and using their statistical calculations for practical applications of quantum physics to systems of large numbers and events.

I strongly suspect that if Einstein knew that quantum physics would enable us to rediscover God and that the quantum God is not benign, he would be very happy indeed.

THE REALM OF THE ARCHETYPES: THE SUPRAMENTAL CONTEXTS OF INTUITIVE EXPERIENCES

Where do physical laws originate? Some philosophers think physical laws are mind-made descriptions of the behavior of physical objects.

Often this is summarized by the question, "Can Newton's law of gravity make even a leaf fall from a tree?" The physicist John Wheeler, in a discussion with two other physicists, approached the question in this way:

> Imagine that we take the carpet up in this room, and lay down on the floor a big sheet of paper and rule it off in one-foot squares. Then I get down and write in one square my best set of equations for the universe, and you get down and write yours, and we get the people we respect the most to write down their equations, till we have all the squares filled. We've worked our way to the door of the room. We have our magic wand and give the command to those equations to put on wings and fly. Not one of them will fly. Yet there is some magic in this universe of ours, so that with the birds and the flowers and the trees and the sky it flies. What compelling feature about the equations that are behind the universe is there that makes them put on wings and fly? (Quoted in Peat, 1987).

The point is that the equations we mentally compose to represent the laws don't fly, but what about the "real" laws behind them, the laws for which the equations are the mental representations, the laws that we intuit and mentally represent as best as we can with our equations? They must fly; they must be potent. Our equations evolve with time; the representations get better and better. But the real laws toward which our mental representations evolve are eternal.

It is a fact that the law of gravity is not a program encoded within a piece of rock that guides the rock's attraction toward the earth. Nor is the falling movement of the rock the result of a program written into its body. There must be an archetype (to use Plato's term) behind the law of gravity that manifests a causal force of attraction between the rock and the earth. And similarly there must be another archetype behind the falling movement of the rock under the earth's gravity. These archetypes must constitute the most esoteric compartment of the possibilities of becoming for consciousness or Godhead—the supramental compartment.

Where does mathematics originate? Mathematics is a meaning given to symbols that represent things, usually physical. So mathematics must come from the mind. And then there are the laws of mathematics. The famous incompleteness theorem proved by Kurt Gödel—a sufficiently elaborate mathematical system is either incomplete or inconsistent—is an example. (This theorem is also notable for its use of tangled hierarchies of logic.) These laws of mathematics must also have an archetypal origin (meta-mathematics).

In biology, there are biological functions—waste elimination, reproduction, maintenance, to name a few—that represent purposive ideals toward which the vital blueprints of these functions evolve. As these evolving blueprints find representation in the physical, biological form evolves purposively toward more complexity.

We can see that there should be archetypes in the supramental that guide the purposive movement of the vital blueprint. Should we be able to make mental mathematical representations of these laws? We should. Some progress may already have been made in this regard (Thom, 1975). This is an area where new research is needed.

There are also archetypes that represent the mental movement of meaning—love, beauty, justice, etc.—that Plato was one of the first to elucidate. These archetypes guide the movement of mental meaning toward a purpose. Can we ever find mathematical representation of the laws of movement of mental meaning? Mathematics itself consists of symbols of which mind gives meaning. To discover mathematical representation of the archetypes for the movement of meaning itself will be a mind-boggling endeavor, but it must be possible.

One thing we already know. The archetypes of physical forces and their vital and mental interaction, the archetypes behind all laws of movement in general, must guide only the movement of possibilities of consciousness. In other words, all movement—physical, vital, mental—is quantum movement. Only consciousness can make a movement manifest through the action of downward causation of conscious choice.

The evidence of the quantum movement of the physical suggests what to search for in the vital and mental movement, as experimental proof that those movements are also quantum. The signatures of the quantum vital and mental realms consist of discontinuity and nonlocality, for which there is ample evidence.

Notice that, in the ultimate reckoning, even the supramental archetypes are quantum possibilities for consciousness to choose from. As mentioned earlier, the experience of such choice is what we call intuitions and creative insights.

Are there "super archetypes" behind the movement of archetypes? We do not know, and we cannot know at the present mental stage of our being.

What is the experimental "proof" of this archetypal, supramental dimension of consciousness? We have already discussed one: the existence and theorizing and experimental verification of the laws of physics. One signature of the supramental is that the elements of this dimension are universal. The universality of the biological laws of behavior of morphogenetic fields would be another proof. But since all earthly life originated from that one first living cell, the geographical universality of biological forms does not prove the universality of the morphogenetic fields. So we could verify this if extraterrestrial life is ever found. Fortunately, our minds did not arise from a common origin, so the universality of the mental archetypes is experimental proof for the universality of some of our dream symbology (of the "big dreams," to use Jung's terminology: Jung, 1971).

CREATIVITY

Creativity is the discovery of new meaning of value (Amabile, 1990). New meaning can be discovered in a new context—this is *fundamental* creativity. New meaning can be invented in an old known context or a combination of old contexts—this is *situational* creativity. Picasso's discovery of cubist art is fundamental creativity; the invention of the fast Internet processor, Google, is a great example of situational creativity.

Where do contexts of profound meaning arise? They are derived from the supramental domain, the archetypes. So the many instances of fundamental creativity in science, arts, music, architecture, mathematics, etc. give us the most definitive evidence of the supramental domain of archetypes.

There are also many reported instances of "inner" creativity or spiritual enlightenment in which the context shift of meaning pertains to one's own self. The creative leap in these cases is the discovery of the true nature of the self, the quantum self, or what Jung would call the *Self archetype*.

Since creativity presents major evidence of both the supramental domain of reality and a quantum signature of the divine, I will present further details in chapter 17.

QUANTUM HEALING

In chapter 12 I mentioned mind-body disease—how faulty meaning processing in emotional situations produces stress, which can lead to disease. How do we heal such disease? We can, of course, deal with the physical level first. But there is plenty of evidence showing that if the faulty meaning processing persists, the disease relapses. Thus we have the idea of mind-body healing—correcting the faulty meaning processing of the mind to heal the diseased body.

But how do we correct faulty processing of meaning? By finding a new context of thinking, right? There is a similarity with inner creativity here. In inner creativity, we find that our inner belief systems or contexts of thinking cannot be changed in a continuous fashion by reading or through discussions with a teacher. Similarly, one has to take a quantum leap to the supramental level of being in order to change the context of meaning. The shift in context for the processing of meaning must arrive discontinuously in order to be effective; in other words, a direct influence of the supramental is essential. And nowhere is the discontinuous nature of a mental contextual shift more spectacular than in spontaneous healing without medical intervention.

Indeed, there exists a large repertoire of cases of spontaneous healing (O'Regan, 1987, 1997), practically instantaneous healing without medical intervention. Many of these cases have involved the overnight disappearance of cancerous tumors.

The physician Deepak Chopra (1990) was the first to suggest the term *quantum healing* to refer to these cases of spontaneous healing. Quantum healing, according to Chopra, is taking a quantum leap to heal oneself. We can clarify this further by saying that the quantum leap is from the ordinary thinking mind of conditioned contexts to the supramental domain to discover a new context for processing meaning.

An example (Weil, 1983) will make this clear. A woman had Hodgkin's disease but refused radiation treatment or chemotherapy since she was pregnant. Her physician suggested an LSD trip, which she took while being guided by her doctor, in order to deeply communicate with the fetus in her womb. When the physician asked her if she had the right to cut off the new life, she felt a communication from it. At that moment she also experienced a sudden insight—that *she* had the choice to live or die. This change in the context of her thinking took some time to manifest in her life, but she was healed. Her unborn child survived, too.

This patient's insight obviously was about her deep self—the suppressed archetype of the quantum self. In this way, quantum healing provides us with direct evidence of the supramental archetypes.

And behind the quantum self, who is the real healer, the chooser of the healing intention? It is quantum consciousness, of course, God. So quantum healing is also direct evidence for God's downward causation. This will be further elaborated in chapter 19.

A physician (allopathic, of course) went to heaven and found a big line at the pearly gates. Being an American doctor, he was not used to waiting in line, so he went straight to St. Peter, the gatekeeper in charge of admission to heaven. Upon hearing his complaint, St. Peter shook his head. "Sorry, Doc. In heaven, even doctors have to wait in line to get in." But just then, one fellow in a white physician's robe with a stethoscope hanging from his neck went running through the gate, paying no heed to the line.

"Ha," said our doctor. "That doctor went in without waiting in line! How do you explain that?"

"Oh," chuckled St. Peter. "That's God. He is returning from a quantum healing episode."

DEPTH PSYCHOLOGY OR HEIGHT PSYCHOLOGY?
IS THE SUPRAMENTAL "DOWN" OR "UP" FROM US?

Much definitive evidence for the supramental archetypes comes from the data on inner creativity, the transformational journey of people. This is a field of study of two recent branches of psychology: depth psychology and height psychology, also called transpersonal psychology. Actually, transpersonal psychology incorporates the wisdom of ancient esoteric traditions such as Indian yoga psychology (Krishnamurthy, 2008).

But psychologists are torn between Freud and Jung's conceptualization of the unconscious, which is the basis of depth psychology, and the concept of the "superconscious," which is the basis of yoga and transpersonal psychology.

As we have seen in the vision of depth psychology, the archetypes of our creative transformational journey rest in the depth of our collective unconscious. We have to delve deep to discover these archetypes, let the unconscious processing take place, and allow what comes up to be integrated. Then we shall arrive in the promised land.

Yoga and transpersonal psychologists view this a little differently. They, too, see the conditioned behavior of the human being as the play of the ego, the domain of behavioral psychology. But they claim that human behavior does not have to stop there, with the development of the conditioned ego. The development can continue beyond the ego, using similar developmental processes but now exploring further dimensions of human potential. There are the ordinary states of our conscious ego, no doubt; we live there most of the time. But we also have momentary experiences of nonordinary, "higher" states of consciousness (intuition). We can cultivate these higher states of consciousness through various conscious techniques such as meditation, one reason that this psychology is also called "height" psychology. Eventually we end up in superconscious states of *Samadhi* (the Sanskrit term for peak experiences of primary awareness, in which the consciousness of the subject tends to becomes one with the object experienced) that have transformative effects. Reaching these superconscious states opens the doorway to spiritual enlightenment that leads to transformation.

In this vision, human development is seen as a ladder that we climb: from the preconscious states of a child, to the conditioned states of the conscious ego, to the superconscious states of the enlightened sage. This is height psychology, which has the further advantage of using a terminology and conceptual framework from esoteric spiritual traditions (as in yoga psychology).

So what is the difference? And is one path better than the other? Controversies and much confusion exist because both tracks hitherto have lacked dynamical foundation. In chapter 6, I outlined the quantum conceptual foundation of depth psychology. Is there a similar foundation for transpersonal psychology using quantum science within consciousness?

CONCEPTUAL FOUNDATION OF TRANSPERSONAL PSYCHOLOGY

The conceptual problem is to explain the ladder of proposed human development: from pre-ego to ego and then beyond ego (transpersonal self), with any other, intermediate homeostatic stages.

The philosopher Ken Wilber (Visser, 2003) begins the explanation with the "great chain of being" of esoteric traditions—body, mind, soul, and spirit. He looks at developmental stages as a progressive climbing of the ladder defined by the great chain. At the first, pre-ego level, the being is entirely physical. Then it goes through the next stage, incorporating the development of the mental ego. But the development does not stop there. It naturally continues to transpersonal stages through the development of the soul level of being. It ends at the highest stage, where it becomes identical with the spirit. At each stage, the being is called a *holon*, meaning whole unto itself and part of some other greater whole. Each holon stage integrates the previous stage and also has something entirely new to offer.

You can see in the great chain of being the five bodies of consciousness if you include the vital energy body: physical body, vital body, mind, soul or supramental body, and spirit (the ground of being). Let us consider the ladder in quantum terms. At each stage, consciousness identifies with what is available for manifestation, for collapse. So at the physical-vital stage, the identity encompasses the physical and vital body; this is the pre-ego (pre-mental) stage (however, see below). Then, representing the mind in the brain begins with cognitive development, greatly facilitated by a language-processing capacity, ending with consciousness identifying with a mental ego. At the next stage, the soul-level learning is explored and consciousness identifies with the transpersonal stages of development, of which Wilber has quite a few. These stages are characterized by "peak" experiences of superconscious states and transformation.

You can call each stage of identity of consciousness a holon if you wish, but there are subtleties, as we'll see later.

This conceptualization looks very different from depth psychology until you recognize the obvious: at the pre-ego level, the mental states belong to the (collective) unconscious, and in the ego stage, the soul states belong to the unconscious. So at each stage, we can consider that we are exploring the unconscious (making it conscious), dipping into its depth instead of climbing the ladder.

The difference of emphasis between these two tracks of the psychology of development becomes obvious when we consider the process in which development actually takes place. As the psychologist Jean Piaget (1977) discovered in child development, the next stage always consists of a creative quantum leap, a discontinuous collapse of new contexts of living, be it of the vital, the mental, or the supramental soul. However, creativity also involves a process consisting of preparation, unconscious processing, sudden insight (quantum leap), and manifestation (Wallas, 1926). The depth psychologists emphasize the unconscious processing, not dwelling on the rest of the (inner) creative process. The transpersonalists emphasize the conscious part of the creative process, preparation and insight, not mentioning the unconscious processing. However, the end goals of both schools—individuation and enlightenment (that implies transformation)—are quite similar.

But of course, all of the stages of creativity are important. In its own way, the difference in emphasis between the two schools has been productive. Whereas transpersonal psychology has helped to legitimize the ancient wisdom paths to God-consciousness, depth psychology has helped chart a relatively new path for the modern human. Both have value in the pursuit of human potential fulfillment. Likewise, both have therapeutic value for helping people in need.

Science is monolithic in the gross material domain, but we should not make the mistake of declaring it a general rule and expect that there should be one science for the subtle. Elsewhere (Goswami, 2004), I have argued in favor of many approaches to subtle-body medicine. Here we should welcome, following cultural anthropologists, different psychologies for investigating subtler aspects of consciousness. Is God deep down or higher up from us? It does not matter what path we follow or how we picture our path.

There are some important controversies, however, that have remained unresolved so far in how the two schools approach human development. We will return to these questions and give a quantum resolution later in the chapter.

We must also note in passing that the materialist model of psychology stops at psychosocial conditioning of the machine that thinks of himself or herself as a conscious ego, because of some apparent emergent epiphenomena, such as consciousness (subjective experiences), free will, etc. There is only mechanical cognitive development in this model; development is then a matter of quantity of knowledge or information, such as the programming of a computer with time. There is no room for human creativity in this model, nor is there any scope for discovering the soul level of values and wisdom. In other words, materialists deny inner creativity. Avowed materialists, such as the philosopher Daniel Dennett, are supposed to be born and live as zombies, collecting information, and then are supposed to die as zombies. And they do live their avowed life as zombies to a surprising extent, as far as an outsider can see. Such is the sad fate of materialists, ironically created by choosing to say "no" to subtler experiences of consciousness.

Is the idea of transpersonal stages of being, or individuation, or what we popularly call *enlightenment* (which implies transformation), or whatever you may call it, empirically valid? If it is, then this is another one of the impossible problems of the materialist view of the world.

Do We Ever Transform?

This is the million-dollar question. Neurophysiologists have their molecules of happiness, endorphins, but even they know that a limited endorphin supply cannot provide the key to lock up all the sites of unhappiness. The spiritual concept of transformation is of a 100-percent-happy person: always equanimous, creative as needed, unconditionally loving to everyone, joy bubbling over, and so peaceful that if you sit near that person for a while, all your restlessness simmers down and you become peaceful. Can a human being be like that? Impossible, say the materialists. Very possible, say the traditionalists; it has happened quite a few times in human history. The founders of the world's great religions are supposed to have been such people. And there still are such people, traditionalists insist.

There are believers, of course; religious fundamentalists still far exceed materialist fundamentalists in number. But if you are a reasonable person and if you look at the spiritual scene without prejudice, doubts may enter your mind.

First, it is easy to find only talkers, teachers who can inspire. Inspiration is important, of course, but you wonder. Does the teacher live the way she inspires us to live? Even in California, the New Age mecca, such skepticism led to the popularity of the dictum, "Walk your talk."

Second, there are those persistent scandals. Sooner or later, scandals seem to engulf all public teachers of spirituality. There are scandals about the misuse of sex, power, money—all the things that cause ordinary people trouble. But aren't we talking about enlightened people? They are supposed to be different, no? The defenders raise their own slogan, "Birds do it, bees do it, and gurus do it, too." Perhaps we should not be naïve enough to believe that enlightened transformation is useful in ridding us of our instincts!

The need for a middle ground should be obvious. But it is still disaster, an impossible problem for materialist science. Can one be 80 percent transformed, or even 60 percent? Does that count? Yes, it does count. Society needs people who are mostly happy, creative, and inspiring, mostly peaceful and wise, and mostly optimistic and loving. These people are mostly environmentally independent, have a sense of humor, do not take themselves seriously, and so accept their imperfection. When a society has an abundance of such people, that society thrives. The opposite happens when there is a dearth of such people.

Here is the good news. The psychologist Abraham Maslow (1968), the founder of the transpersonal psychology movement in America, collected conclusive data indicating that people can be divided into three mental health categories: normal, pathological, and positive. About 5 percent of all people have positive mental health, compared with about 30 percent for pathological cases and 65 percent who are normal.

The people of positive mental health that Maslow studied also had frequent peak experiences—another name for the quantum leap to the supramental. A wonderful confirmation of Maslow's data on people who take quantum leaps has come from the data on near-death experiences. Cardiac surgery can sometimes restore "clinically dead" people to life. Some of these people describe astounding "peak" experiences while in near-death coma. The psychologist Kenneth Ring (1984) did an exhaustive study of these people and found that many of them are (partially) transformed and live a life of positive mental health.

Yes, there is a God, because maybe as many as 5 percent of the people on earth have positive mental health; they are optimistic, loving, environmentally independent, creative, humorous, etc., most of the time. These people, in the language of the new science, live in God-consciousness at least sporadically.

So the idea of *samadhi*—creative insights of primary awareness, followed by transformation or individuation—is valid, except that the idea of 100 percent transformation has to be considered more cautiously.

THE PRE/TRANS FALLACY

I would like to offer a resolution of the much-touted *pre/trans fallacy* that is a prime example of the confusion within the developing new paradigm of psychology. The transpersonalist Ken Wilber does not seem to agree with the depth psychologist Carl Jung's ideas of human development. According to Jung, the early child lives as one with the archetypal (quantum) Self, as a child of the divine. With ego development, the Self is repressed. And then post-ego development recovers the repressed Self and restores it to the foreground. For Wilber, the self of the early child is limited to only a physical body identity. And although in Wilber's scheme a person at any stage of development can have experiences of the self in another stage as transcendental experiences, in actuality this access is quite limited. Wilber's concept of the holon says that for a child the experience of a later holon stage, like the soul or supramental with rich archetypal content, is almost impossible. This is because the child has no way to manifest such an experience or process such an experience. That experience requires an ego.

Here's how Wilber (2001) expresses his idea:

> The essence of the pre/trans fallacy is itself fairly simple: since both prerational states and transrational states are, in their own ways, nonrational, they appear similar or even identical to the untutored eye. And once pre and trans are confused, then one of two fallacies occurs.
>
> In the first, all higher and transrational states are *reduced* to lower and prerational states. Genuine mystical or contemplative experiences, for example, are seen as a regression or throwback to infantile states.... In these reductionistic accounts, rationality is the great and final omega point of individual and collective development, the high-water mark of all evolution. No deeper or wider or higher context is thought to exist. Thus, life is to be lived either rationally or neurotically.... Since no higher context is thought to be real, or to actually exist, then whenever any genuinely transrational occasion occurs, it is immediately explained as a *regression* to prerational structures. ... The superconscious is reduced to the subconscious, the transpersonal is collapsed to the prepersonal, the emergence of the higher is reinterpreted as an irruption from the lower....
>
> On the other hand, if one is sympathetic with higher or mysti-

cal states, but one still *confuses* pre and trans, then one will *elevate* all prerational states to some sort of transrational glory. . . .

In the elevationist position, the transpersonal and transrational mystical union is seen as the ultimate omega point, and since egoic-rationality does indeed tend to deny this higher state, then egoic-rationality is pictured as the *low point* of human possibilities, as a debasement, as the cause of sin and separation and alienation.

Freud was a reductionist, Jung an elevationist—the two sides of the pre/trans fallacy. And the point is that they are *both* half right and half wrong. A good deal of neurosis is indeed a fixation/regression to prerational states, states that are not to be glorified. On the other hand, mystical states do indeed exist, beyond (not beneath) rationality, and those states are not to be reduced.

So the pre/trans fallacy. The soul level can be developed only after the ego development. The development of the soul level is not a regression to the childhood.

Thinking the quantum way will let you see through the problem to the other side. Freudians are wrong, no doubt, but there is no need to make Jungians wrong. At each stage, there is a conditioned identity of consciousness and a creative identity—the quantum self, Holy Spirit-consciousness. Sure, Wilber is correct: initially the baby's identity is primarily with the physical/vital body. But the baby's mental unconscious processing is done without conditioning, without any ego; it always processes it in God-consciousness. When conscious choice takes place, the result is an immediacy of experience that we call the *quantum self or Holy Spirit experience*. This is why it is not incorrect to say that the early child lives much of his time in God-consciousness, not with conscious wakefulness, but unconsciously. It is quite right that Hindus regard children as God until they reach age five.

But Jungians get confused and caught up in their own language, too. As ego develops, the quantum self is harder to reach because quantum leaps are more difficult to take. The quantum self does not disappear; there is no regression. Nor does the ego push it away. It is simply the nature of conditioning that creativity is more difficult when we have lots of memories. And yet, as Wilber says, memories should not be looked upon as detrimental to later development. A child has easy access to quantum self experiences, but is unable to make mental representations of these experiences. Precisely because our adult egos have this vast range of sophisticated material, we can manifest and make representations of creative insights that require such sophistication. Otherwise, we would be rediscovering the wheel over and over.

ALTRUISTIC BEHAVIOR

Altruistic behavior undoubtedly exists. Many people in all cultures often give a helping hand to others in need without demanding anything in return. Where does altruistic unselfish behavior originate? The conceptual schema that tries to incorporate altruistic behavior into our usual norm is called *ethics*.

Of course spiritual traditions make ethics more complicated than just the conceptual context for the study of altruistic behavior. In most spiritual traditions, for example, ethics is about discriminating between good and evil. We humans have a discriminative function called *conscience*; we suffer pangs of conscience if we fail to choose good. Thus we have the simple statement of spiritual ethics, "Be good, do good" (to yourself and others), from the Hindu Swami Sivananda. Another statement, this one by the Rabbi Hillel, expresses the same concept:

If I am not for myself, who am I?

If I am only for myself, what am I?

And still another statement, this one from Christianity: "Do unto others as you want others to do unto you."

It is this discriminative conscience that enables us to do good. Where does conscience originate? It is the bidding of the supramental or soul level of being. In this way, our altruistic behavior proves the existence and reality of the supramental domain.

Ethics is important for spiritual traditions, because being good is a godly quality; it is a virtue. If you acquire it, it takes you closer to God. If you shun it or do evil, that behavior takes you away from God.

Religions like popular Christianity put it more bluntly: if you are virtuous, you go to heaven when you die, and if you are sinful, you go to hell after death. (For a crude but hilarious depiction of the latter, see the movie *Ghost*.)

It is this latter depiction that does not appeal to some modern people. But what if the religions are right? Is ethics compulsory? Suppose ethics is a science and is compulsory like scientific laws—what then?

The philosopher Immanuel Kant sided with religion and believed that ethics is the categorical imperative, which he expressed succinctly in *Grounding for the Metaphysics of Morals*: "Act only according to that maxim whereby you can at the same time will that it should become a universal law." It is an inner moral law for each of us and is compulsory. The imperative arises because we have a moral sense of duty or duties that we can figure out by reasoning. And oh, yes, for Kant, the inner moral law came from an immortal soul, another name for the supramental. So for Kant, altruistic behavior was imperative and it proved the soul or supramental level of our being.

But obviously ethical law, or inner moral law if it is that, cannot be compulsory in the same cause-and-effect sense of science. If you try to violate the law of gravity by trying to fly, you fail: you cause an effect now. If you cheat ethics and get away with it, where is the failure? What is the effect that you cause? None is apparent unless you take hell seriously—and that's later, not now!

Well, you suffer from the pangs of conscience, you may think. But is conscience real for everyone? In Fyodor Dostoyevsky's classic novel *The Brothers Karamazov*, the two brothers Ivan and Alexei are obsessively torn between right and wrong, good and evil. But the novel was published in 1880; that was another era. Can you imagine people of our time similarly perturbed by the ideas of good and evil, right and wrong?

But altruism is real, empirically proven behavior, and not compulsory for everyone. A substantial number of people help out others unselfishly, so altruistic behavior must be proving something. But what?

Biologists have tried to answer this question with the idea of the selfish gene (Dawkins, 1976). According to this thinking, we are gene machines, the way our genes propagate and perpetuate themselves. Consistent with that purpose, our genes ensure that we behave altruistically toward those people with whom we share some of our genes. For example, we will tend to be altruistic to our own children or parents, but proportionately less so toward cousins, and much less altruistic to the cousins' children.

This idea is interesting, but easily disputed by the vast amount of data (anecdotal, to be sure) of saintly people giving unselfish help to completely unrelated people without expecting anything in return. Mother Teresa is only one glaring recent example.

So, again, what does altruism really prove?

ETHICS IN THE CONTEXT OF SCIENCE WITHIN CONSCIOUSNESS

With the idea of nonlocal consciousness, ethics is easily validated and altruism is easily explained. If you and I are not separate, if we both belong to the same nonlocal consciousness at a deeper level, then certainly I may feel an urge to give you a helping hand when you are in need and vice versa. We are just helping ourselves! Altruistic behavior, indeed any ethical behavior, comes from the urging of our nonlocal self-archetype or rather its mental representation (call it conscience). It proves the supramental level of being, the soul.

We must note, however, that there is a predominantly vital component to our conscience; it is a "heart" thing. People who are more sensitive to vital energy, people with open hearts, suffer more from pangs of conscience than people who are less sensitive to vital energy, people with primarily thinking minds.

Early conditioning further complicates any discussion of conscience. For example, religious fundamentalists often have a strong sense of ethics and morality, but it is mostly made up of conditioned beliefs. When there is subtle complexity in making an ethical choice for right action, such as extending help to people beyond one's own "clan," the conditioned conscience may not be able to resolve the ethical dilemma. One may need to take a quantum leap to the supramental to get a clear insight about ethical action. But if the conditioning is substantial, such quantum leaps are unlikely to happen.

And of course, altruism is not compulsory. If we are not feeling energy-sensitive, if the situation is not black and white, a conditioned conscience most likely will not hear the intuitions of the self-archetype for ethical action.

As you can see, the new science gives us the proper context for understanding all the facets of altruistic behavior, and it proves the existence of our supramental level of being, the soul.

Chapter 14

Dream Evidence

During the period from June 1998 through 2000, I was a senior scholar in residence at the Institute of Noetic Sciences in the San Francisco Bay Area. There I had a very enthusiastic research assistant, Laurie Simpkinson. She needed a project for research. Although I am a theoretician, I saw an opportunity there. I have always been interested in dreams and have done much analytical work with my own dreams. When I found out that Laurie shared a similar interest, I chose dreams as her research subject. Naturally, for collecting data, we set up a dream group at the Institute. Most of what is reported in this chapter is the result of our collaboration.

Most of our knowledge about the science of dreams comes from two sources: neurophysiology and psychology.

Neurophysiologists tell us, for example, that dreams mainly happen during REM (rapid eye movement) sleep that has a specific brain-wave signature. Neurophysiologists also make a good case that we make our dream pictures from the Rorschach of white noise that the electromagnetic activities of the brain provide (Hobson, 1988). However, neurophysiology is a materialist ontology where the meaning of dreams can never be reached. In this absence of a complete theoretical framework, the specter of dualism hangs over the neurophysiological picture.

Psychologists, beginning with Freud and Jung, have discovered a great amount of therapeutic value in dream analysis with their clients, because of the rich meaning embedded in dreams. According to Jung, dreams tell us about the great myths that run through our lives. Many others believe that dreams help formulate and perpetuate personal myths that we create and that we live by.

But why should dreams carry such deep meaning? Many scientists are openly skeptical, insisting that dreams are "nonsense" and "without meaning." Some scientists go further, claiming that dream analysis may be detrimental to our mental health. Biologists Francis Crick and Graeme Mitchison (1983) write, "We dream in order to forget." (Later, Crick and Mitchison (1986) revised their position slightly: "We dream to reduce fantasy and obsession"—dreams are a way to forget things that might otherwise intrude in our lives.) They explain (1983), "Attempting to remember one's dreams should perhaps not be encouraged, because such remembering may help to retain patterns of thought which are better forgotten. These are the very patterns the organism was attempting to damp down." Nevertheless, our fascination with dreams remains, because there is evidence not only of their therapeutic importance, but also of their importance in creativity.

The undeniable fact remains: we dream. But why? What function do dreams perform? How should we go about understanding them? What are they proving to us?

Although there is agreement that dreaming is a state of consciousness just like waking, there are philosophical problems with taking dreams seriously—or at least as seriously as our waking experiences. One issue is continuity. We take our waking life seriously because there is an ongoing continuous character to it. The same objects appear repeatedly; we wake up from a dream and find ourselves in the same bed in the same room where we went to sleep. Also, a cause-effect connection is clear between events of our waking experience. Dreams, on the other hand, seem to have no continuity: you dream, wake up, and go back to sleep and dream, but ordinarily you won't return to your previous dream scene. Seldom can one find any cause-effect relationship between dream episodes. So, how can we take dreams to be real in the same sense that we consider our waking life to be real?

In contrast to this way of philosophizing, the mystics of the world take an opposite view. They agree that dreams are unreal, but they claim that our waking life is also a dream and unreal in a sense. Dreams are the creation of the "little me" and the waking life is the dream of the "big dreamer," or God within us. Mystics say that when we realize that there is no difference between waking and dreaming, that they are just different states of consciousness with similar values, then our perspective of living shifts to God-consciousness and we become liberated from the shackles of worldly boundaries.

The mystics' point is at least somewhat corroborated by the recent discovery of lucid dreams (LaBerge, 1985), in which we are aware that we are dreaming and have the ability to guide the dream to reveal solutions to problems in our waking lives. This raises the question, if we are sufficiently awake to realize that we are dreaming, why can't we realize we are dreaming while we are awake?

Then there are data about telepathic dreams and precognitive dreams that further complicate our attitude toward dreams. If dreams can tell us about "real" physical events distant in space and time, how can we not take dreams seriously?

MATERIALISM OR SUBTLE BODIES?

The earlier mentioned neurophysiological models of dreams are only able to respond to questions relating to the physical data (EEG report) gathered from measuring brain activity. Since physical matter for materialists is the ground of all being, the brain activity measured on the EEG report is the final and only reality. The person at home experiencing the dream sensations, feelings, and thoughts that correlate with the brain state is secondary to the physical brain state, as are the experiences. In materialism, consciousness is either an epiphenomenon of matter (brain) or (implicitly) a dual body. If you subscribe to epiphenomenalism, all explanations consist of finding a deeper, "objective" explanatory realm; such an explanation makes the question of the subject of an experience a "hard question" (Chalmers, 1995).

Furthermore, if epiphenomenalism is the case, then meaning cannot be accounted for, because with finite physical symbol processors such as the brain, one cannot arrive at meaning. Who makes meaningful images out of brain noise? There is no homunculus sitting in the brain looking at a TV screen. Something, a subtle body—namely, mind—must lie outside the material world to establish meaning.

The materialist model also fails at explaining telepathic and precognitive dreams, because such nonlocal qualities cannot be explained in materialist science, where locality reigns supreme.

In the materialist view, since dreams are epiphenomena of the brain, there is no causal potency in them, let alone causal potency as strong as the waking state. Therefore, if we are to understand the meaning of dreams and their nonlocality and causal potency, we must look outside the materialist view.

To introduce a proper science of dreams, we must consider consciousness as the ground of all being, consisting of five levels or worlds of being: the physical, the vital, the mental, the supramental, and the bliss ground of being. The most important objective of this chapter is to prove the veracity of a fivefold classification of dreams, thereby establishing the validity of our five bodies in consciousness.

What do dreams prove to us, then? They give quite definitive evidence that we are not one material body, but five bodies within consciousness.

WHO DREAMS? THE ANSWER FROM QUANTUM PHYSICS

Who dreams? This question presents a conceptual quandary in materialist thinking, because an objective explanation of a subjective experience (of the dreamer) is an unsolvable paradox. Quantum physics gives us the way out.

Who dreams? Consciousness dreams by converting waves of possibility into the actual events of the dreams and in the process dividing itself into two parts: one part, the dreamer, sees itself separate from the other part, the dream objects.

One reminder. How is the ego-individuality created? The answer, to repeat, is that experiences create memory; this memory feedback modifies the dynamic of quantum movement in favor of our past responses to stimuli. In other words, we become conditioned to respond in a certain way, albeit an individualized way, rather than retain all the freedom we have when we are naïve. These conditioned patterns are what creates our individual ego, along with the history contained in the memory.

Thus, it would be incorrect to assume that the continuity of a body over time comes from the actual world. Rather, the continuity in the actual world is an effect caused by the conditioned way we experience it.

THE STATES OF CONSCIOUSNESS

Human experience corroborated by brain-wave data allows us to enunciate the three types of states of consciousness:

1. **The waking state.** Here we have both external and internal awareness.
2. **The dreaming state.** Here, only the internal state of awareness exists.
3. **The deep sleep state and other *nirvikalpa* states.** Here there is no subject-object awareness at all, no collapse of the possibility waves. The Sanskrit word *nirvikalpa* means "no separateness."

Can we say that the waking state is any more real than the dream state, just because when we are awake we have both external and internal awareness? We should not be too hasty. There are data suggesting that occasionally even in dreams, we have objective (therefore external) awareness.

In yoga psychology, consciousness has three defining aspects: existence, awareness, and bliss. We can easily see that these qualities are equally available in both waking and dream states. The case for existence and awareness is obvious, but it is valid even for bliss. Just as we can enhance the bliss level of our waking lives through spiritual practices and quantum leaps, similarly we can also enhance the bliss level of our dream time through methods developed in esoteric traditions called *dream yoga*—practice of awareness while dreaming.

But how about the philosopher's contention that dreams do not have any cause-effect continuity, that they jump around from episode to episode without any apparent causal continuity whatsoever? In contrast to the apparent fixity of waking awareness (needed for a reference point to communicate with others), where quantum uncertainty is camou-flaged, dreams retain their quantum nature to a much larger extent, only yielding somewhat to Newtonian fixity because of conditioning. So in dreams, we have a conditioned continuity; this gives us the story line of a particular dream episode. But when the episode changes, we have the opportunity to experience the causal discontinuity of quantum collapse. In truth, however, very often there is a subtle continuity even in episodic changes. But you have to look at the meaning to find it.

This brings us to the other question that philosophers ask. When we wake up from a dream, we return to the same waking reality (perhaps with minor changes that are easy to explain), but when we go back to dreaming, we seldom encounter the same dream reality. So how can dream reality be taken seriously? The answer to this question is that dreams speak to us about the psyche—their concerns are feeling, mean-ing, and contexts of meaning. So we have to look for continuity not in content, but in meaning and feeling. When we do that, we can readily see that most often, especially during the same night, we do return to the same dream reality in terms of meaning or feeling. The contents and images change, but the associated feelings and meanings remain the same.

This way of looking at dreams can also resolve another question that people sometimes ask. In our waking state, we can and do talk about our dreams. Why can't we similarly talk about our waking life while dreaming? But we do! Except the language of dreams is made up of feeling, meaning, and the contexts of meaning (archetypal symbols). This language is a little hard to penetrate. When we do penetrate the language, we discover that in our dreams we do indeed speak about the problems of our waking life, that we reenact them in this way and that way and sometimes even find creative solutions.

So psychotherapists who encourage their clients to engage in dreamwork mainly at the meaning level are helpful. It is good to see that, beginning with Freud (1953) and then Jung (1971), Adler (1938), and others, psychotherapists assume that the meaning the dreamer sees in his or her dream symbols is most significant. The gestalt psychologist Fritz Perls (1969) summarizes this attitude best when he says, "All the different parts of the dream are yourself, a projection of yourself."

The new quantum science of dreams agrees: a dream symbol is a projection of yourself to the extent that it represents only the personal meaning that you attribute to that symbol in the overall context of the dream, with proper attention given to the feeling aspect. Especially important are the other human characters in your dream. When you see your mother in a dream, she is you, or that part of you which mirrors your perception of her. Of course, there are also universal contextual symbols (that Jung called *archetypes*), representing universal themes that appear in dreams in which we universally project the same meaning. One such theme is the "hero's journey," in which the hero goes in search of the great Truth; the hero finds the truth, is transformed, and returns to teach others.

So dream analysis is not only a science but also an art, since one has to look for the personal meaning within the context in which the symbol occurs. Some therapeutic schools suggest taking the dreamer through the feeling experiences that occur during the dream and doing the analysis only when the proper feelings are re-experienced by the dreamer. This is good strategy.

The meaning level of life also plays out in waking events, but we get so sidetracked by the clamor of the fixed symbols in waking life that we seldom pay attention to their meanings. For example, suppose that one day you have an unusual number of encounters with stop signs while driving around town. Would you stop to think that this might be some kind of synchronicity? Dreams give you a second chance. The same night you may dream that you are driving your car and then you come across another stop sign. Upon waking up, you may easily realize that the car is representing your ego and the stop sign is attracting your attention to put a stop to your rampant egotism.

THE NEW CLASSIFICATION OF DREAMS

There are many classification systems for of dreams. One of them is to label the dream by the particular school of thought that most easily explains it: thus, we have Freudian dreams (for example, wish-fulfillment dreams), Jungian dreams (in which archetypal symbols appear), Adlerian dreams (that reveal the dreamer's private belief system, its logic, prejudices, and errors), and so forth. But this kind of classification seems quite arbitrary and ambiguous.

Can the new science within consciousness explored here lead to an unambiguous classification system for dreams? The answer is yes. Most dreams can be better analyzed and understood from the viewpoint of the five bodies—the physical body, the (vital) energy body, the mental body, the supramental theme body, and the bliss body.

1. **Physical body dreams.** These are the so-called *day-residue dreams* from the physical body, in which the memory of the physical is role-playing in the dream.

2. **Vital body dreams.** These are nightmares, in which the dominant quality is a strong emotion, such as fear.

3. **Mental body dreams.** These are dreams in which the meaning of the symbols dominates, rather than the syntax or content, such as pregnancy dreams and flying dreams. Many recurring dreams (not nightmares) also fall into this category. These dreams tell us about our meaning life, the ongoing saga of the mind.

4. **Supramental dreams.** These dreams contain objective universal symbols, the Jungian archetypes. They tell us about the ongoing exploration and unfolding of the archetypal themes of our lives.

5. **Bliss body dreams.** These are rare dreams in which the affairs of the physical, the vital, the mental, and even the supramental are transcended. The dreamer wakes up with a deep sense of bliss, grounded in being. Here the dominant body involved is the unlimited bliss body, the eternal in us.

However, there is one caution. Sometimes, dreams play out simultaneously at more than one level. Sex dreams, for example, have not only physical dream typology, but also that of the energy body—sexual energy. Creative dreams take a problem from the physical life (symbols stand for what they are) and bring archetypal images to suggest a solution.

Let's now illustrate the classification with some examples.

My collaborator in dream research, the psychologist Simpkinson, had a dream that she was in bed and her cat was scratching the carpet. As she was about to get out of bed to stop the cat, she woke to find the cat on top of her, scratching and clawing at the blanket. This is a physical body dream, where the dream of the cat scratching the carpet came from the influence of the cat clawing at the dreamer. Physical body dreams also include those that simply repeat the day's activities, especially those that leave a mark in the muscle memory of the physical body.

Next is an example of a predominantly vital body dream. "Nancy," a member of the Institute of Noetic Sciences dream group, talked about a recurring theme of many of her dreams. As an example, she shared this emotionally charged dream with the group:

> I was walking up the driveway and my sister said she was leaving, and then I walked into the house and nobody was there. I looked and looked in every room and nobody was there—they had all left me. And at the same time it is scary because I feel like there is a ghost or something in the house.

Driving this dream is the emotion of fear—fear of being left alone and fear of the ghosts, etc. From that perspective, the symbolic images are those of the dreamer's psyche (the house), and she is fearful that she'll be left alone with the "ghosts" there. This fear was relevant to her waking life as well, since her lifestyle precluded any opportunities to spend time alone with herself.

After sharing this dream, Nancy went on to tell of an early childhood experience when she was playing with her siblings outside her house. At some point, she ran back inside to quickly change her clothes. Her siblings then played a joke on her and hid so that she would think they had left without her. She remembered looking around the outside of the house, thinking she had been abandoned.

The mental body had taken this early experience as a personal archetype for Nancy, in the sense that when the feeling of isolation came up in her life, it manifested itself with this familiar story. Thus, the recurring dreams around this theme were communicating the sense of isolation in her emotional life that needed attention.

When we look at the dream symbology, we can take the house to mean Nancy's psyche. The fear of ghosts indicates that being alone in the psyche is a scary experience for her. Nancy did spend a lot of time alone, but further investigation revealed that it was usually doing something, such as reading a book or cleaning the house. The issue here was the lack of time spent doing nothing—just being with herself. Both in her waking life and in this dream, there was fear around this idea. This nightmarish aspect is based in the vital body of feelings, which is also the area of the psyche that was demanding attention.

This dream revealed the need for solitude and calm. Two weeks after this dream, Nancy unexpectedly had to find a new living situation and moved into an apartment by herself. However, it wasn't until she mentioned her move at the next group meeting that she was able to connect the story of the dream to the manifestation of her actual situation. Although the move into a solitary place was not the entire solution, as she still needed to use the space to spend time by herself, it was another important symbol suggesting the need for being alone in her psyche. It is very important to see how both her waking life and her dreaming life manifested symbols that were relevant to her personal growth.

Although the following dream (contributed by another member of the Institute dream group, "Julia") has the vital body characteristic of a disturbing emotion, it can be primarily understood as a mental body dream where the meaning of the symbols dominates.

I was on a boat with my husband and sons. We reached the first destination and then the boat started to sink. I went downstairs to where my purse was upturned in the water and tried to collect my belongings. I was angry that my sons and husband didn't seem interested in helping me. The boat had then changed into a canoe. I was very concerned about gath-

ering my belongings because we all needed to catch a plane
that was leaving soon. I finally realized I just couldn't make the
plane, but upset that no one would help me or cared to.

In analyzing this dream, it became apparent that Julia's psyche was
adjusting to newfound solitude as her youngest son was graduating and
leaving home. This dream reflects going down into her psyche (down
into the boat) where none of her other ego identities (her family) are
interested in joining her. There she collects those things that have fallen
out of her purse—her driver's license (literally her ID), photos of family
(how she identifies herself), her wallet (money being a symbol of value),
etc.—in order to pull together who she is. The canoe, which is a self-
powered vehicle, showed that her psyche had changed from the com-
munal boat excursion since now she was on her own. She then realizes
that they will not all be able to continue together, as she "just couldn't
make the plane."

Here's an illustration of the supramental body dream. Simpkinson
had a simple dream the first night out on a vision quest. She dreamed of
walking around the wilderness where she and several others were on
their vision quest. As she met up with them in warm camaraderie, it
began to rain. The rain showered down, washing the entire hillside
where the vision questers were staying.

This was a purification dream, initiating her into the vision quest.
Water, in the form of rain, is the archetype of the unconscious. Since it
was the unconscious she was hoping to learn from during her quest, this
came as a significant blessing. Not only was the rain cleansing her, but
it also was touching her, coming down to where the unconscious and
the quester could meet. In this way, the unconscious was agreeing to be
open and "shower" her with its presence.

The following is an example of a bliss body dream, in this case
arrived at through lucid dreaming (Gillespie, 1986).

If conditions permit me to concentrate for long without [dis-
ruptions in the dream], I gradually lose body awareness and
approach the total elimination of objects of consciousness.
Mental activity ceases. I have reached this point of pure con-

sciousness, but have not held onto it that I know of. Inasmuch as sense awareness and mental activity have ceased, I have transcended my physical and mental self. . . .

The final phenomenon is the fullness of light. . . . It usually appears like the sun moving down from above my head until all I see is brilliant light. I become aware of the presence of God and feel spontaneous great joy. As long as I direct my attention to the light, I gradually lose awareness of my dreamed body.

To lose dream imagery and awareness of myself in the evident presence of God is to experience transcendence of myself. This is the experience, whatever the explanation. Fullness of light, awareness of God, gradual loss of awareness of myself, joy (often called bliss), and uncontrollable devotion are phenomena mentioned commonly in mystical literature.

This example describes the loss of the ego identity as the emergence into light and great joy take over. All previous classifications of dreams dissolve, since there are no constructs of symbolic meaning in the bliss state. There is only pure bliss—absence of separateness.

MORE ON DREAMS AND PSYCHOTHERAPY

Why are dreams useful for psychotherapy? Freud had it right when he realized that there are mental processes going on in our unconscious, but we are not consciously aware of them in our waking life. Memories of trauma are activated in possibility whenever a similar stimulus presents itself, but the repression dynamics prevent us from recalling and manifesting such memories. So these memories affect our actions via unconscious processing and lead to behavior for which we cannot find any rational explanation. This makes us neurotic. In the dream state, the physical component of the ego, the body identity, is missing. This weakens the ego, so the usual ego guard against repressed memories is weak. Therefore, these memories can resurface in dreams. And this is a boon to psychoanalysts and, in fact, psychotherapists in general.

Dreams also tell us about the mental and emotional ego more directly than our waking experiences. Through analyzing dreams, therapists can get a sense of the meaning structure—the mental ego—that is part of each client's personality. And the same is true about the emotional structure—the mentalized vital-energy ego. The job of therapy is often to break up these rigid structures, so knowing about them can be invaluable for the therapist. And this knowledge is quite available in the vital body and mental body dreams.

CREATIVITY IN DREAMS

There is much anecdotal evidence of creative breakthroughs in dreams. The most famous perhaps is August Kekule's dream about snakes gamboling together in a circle, giving him the insight that electron bonding in a benzene molecule is circular, a radically new concept. And Niels Bohr is supposed to have developed his atomic model inspired by a dream. Werner Heisenberg discovered the fundamental equation of quantum mechanics in a dream. And it's not just scientists who get their ideas in dreams: Beethoven got the idea of one for his canons from a dream. There are many such instances from other musicians, artists, writers, and poets (Goswami, 1999).

So why should dreams facilitate creativity? The creative process consists of four stages: preparation, unconscious processing, quantum leap of insight, and manifestation. In waking, we are identified with our bodies and physical stimuli dominate our waking life. In dreaming, the body identity is missing; we are wholly identified with the psyche. As a result, many things that are normally relegated to unconscious processing in our waking experience, we now release and collapse into experience in a dream. We cannot precipitate a physical event. But with the help of the noise/Rorschach available in the brain, we can experiment with making images of our ideas at the level of feeling and meaning and, once in a while, we are rewarded when a quantum leap occurs in the context of meaning upon waking, based on the dreaming expedition.

The English Romantic poet Samuel Coleridge graphically described the dream journey through symbols that helped him compose his famous poem "Kubla Khan" in this way : "What if you slept, and what if in your sleep you dreamed? And what if in your dream you went to heaven and there plucked a strange and beautiful flower? And what if, when you awoke, you had the flower in your hand?" Well said.

THE EQUIPOTENCY OF WAKING AND DREAM STATES OF CONSCIOUSNESS

Now we come to the important question. Are dream states as potent as the waking states? Is our dream life to be taken with equal seriousness (or with equal levity, as mystics do) as our waking life? There are quite a few phenomena, some old and some new, that point to an affirmative answer. Among these phenomena are dream telepathy, precognitive dreams, crossover dreams, shared dreams, and lucid dreaming.

As this chapter suggests, dreams use symbols of the waking world to create not content, but the feeling, meaning, and context of meaning. However, telepathic dreams (the nonlocal transfer of information across space by our nonlocal consciousness collapsing similar experiences in two correlated people), precognitive dreams (nonlocal transfer of information across time), and crossover dreams are exceptions to this general rule. In these dreams, certain objects of waking reality do literally represent and mean those objects. It often happens that a telepathic or precognitive dream predicts the death of a close relative in this way; that is, death in these cases means the death of a physical person and is not a symbol for something else. Thus, in this kind of dream, the external physical world and the internal world of the psyche become equivalent. This suggests that at least these dreams are as "real" as the physical world.

In connection with dream telepathy, the research of psychiatrist Montague Ullman, parapsychologist Stanley Krippner, and psychic and editor of *Psychic*, Alan Vaughan (1973), carried out at Maimonides Hospital in Brooklyn, New York, over a decade is definitive. I will give more detail in chapter 16.

Shared dreams are when two people dream the same basic dream or when they sometimes appear in each other's dreams (Magallon and Shor, 1990). Dreams are ordinarily internal, but if two people share a dream, they are being elevated to a consensus reality through nonlocal correlation. How then can we deny that dream reality is in the same league as waking reality?

The best proof of this equipotency of dream and waking life would consist of finding the answer to this question. We use dreams in our waking life to solve problems of our waking life. Do we similarly, while dreaming, use the material of waking life to solve problems of the dream life? The prediction of the present theory is that we can. This question should be experimentally investigated by engaging the symbols of dreams as "real" objects in your waking life. For example, if you see the recurring symbols of clocks in your dreams, I suggest you engage with physical clocks in your waking life and see what that does to your dreams.

Lucid dreams—in which we are aware that we are dreaming while dreaming—are another vehicle for investigating the equipotency of dreaming and waking life. I mentioned earlier that it is a good hypothesis that in dreams all the characters are in some way the dreamer himself or herself. According to the present theory, it should be possible, with some practice and with creativity, to realize this even within the dream, while lucid dreaming—that the dreamer is privy to the "inside" of all the dream characters. This realization is the mystical realization of oneness of consciousness.

When we realize that we are all that is in the dream reality, that realization should carry through to the waking awareness as well. We see that the waking reality is also a dream created by us and that everything in the waking reality is also us. This answers the mystical question "Is the waking reality really a dream, a dream of God?" affirmatively. Thus this kind of lucid dreaming should be a grand subject of experimental investigation.

In this way, dreams not only give definitive scientific evidence for the subtle bodies, but also have the potency to directly reveal to us the nature of the entire reality.

Chapter 15

Reincarnation: Some of the Best Evidence for the Soul and God

have heard that the Dalai Lama was once asked if there was any scientific research that would ever prompt him to give up his Buddhist beliefs about spirituality. To this the Dalai Lama is supposed to have replied that if scientists could ever prove that reincarnation never happens, he might change his mind.

What is reincarnation? And what makes it such a definitive gauge of spirituality?

Reincarnation is the idea that there is some essence of ourselves that survives death and is reborn in another body. In popular parlance, this essence is called the soul; however, the meaning of the word "soul" is somewhat extended from its use in chapter 13. In the context of reincarnation, soul denotes the entire "subtle body" consisting of the vital, mental, and supramental components. Reincarnation can be readily understood within the model of our expanded selves that we have been exploring in this book. (For details, see Goswami, 2001.)

Who am I? I have a physical body. Additionally, I have a subtle individual vital body defined by my vital habit patterns—the specific ways in which I engage this body. In chapter 11, I showed the yang and yin dominance of the vital body in connection with traditional Chinese medicine. The relative amounts of yang and yin is one way to define my vital individuality. I also have an individual mental body defined by my mental habit pattern. As part of my vital and mental bodies, I also have a repertoire of discovered archetypal contexts of feelings and meanings. And if I live these vital-mental representations of the archetypes, I even have physical representations of them.

You can see that while the physical body is structural, our individual vital and mental bodies are functional. The conglomerate of individualized vital and mental bodies, along with the universal supramental body, is called "soul" in the reincarnational context. To avoid ambiguity, I call it the *quantum monad* (Goswami, 2001).

Because the quantum monad is functional, it is not memory recorded in structure, but a quantum memory that Easterners call *Akashic memory.* (The Sanskrit word *akasha* means nonlocal—beyond time and space.) It is memory akin to the laws of physics: it affects us and it guides our behavior from a transcendent domain. The difference between the physical laws and Akashic nonlocal quantum memory is that the former are universal and the latter tends to be personal.

But the quantum memory does not have to be personal for only one lifetime. If many physical human bodies in many different times and places express the same developing quantum monad, the same quantum memory, they are called reincarnations of one unique quantum monad (figure 15-1). Empirically, it is found that these incarnations or past lives of ours are nonlocally correlated to one another and, under special circumstances, we are able to glean the local memories from each. In fact, the data on such past life recall—akin to mental telepathy across time and space—provide definitive proof of downward causation by nonlocal consciousness (Stevenson, 1973, 1977, 1983).

You now can understand the Dalai Lama's comment on reincarnation. The reason for its supreme importance is that reincarnation data

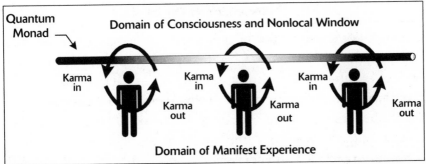

FIGURE 15-1. Model of reincarnation. The quantum monad and the nonlocal window provide the thread that connects our various incarnations over space and time.

proves all the three essential aspects—downward causation, subtle bodies, and supramental godliness—of religions and spiritual traditions in one fell swoop. Reincarnation is impossible if matter is the ground of being; moreover, the data directly prove that our subtle bodies are recycled, confirming their existence. Why reincarnate? Only through many incarnations can we gradually acquire supramental godliness (learned repertoires of supramental archetypes); that is practically impossible to do in one life.

Reincarnational data have a gross aspect and a subtler one. The gross aspect consists of all the reincarnational memories that people recall, not only children (Stevenson, 1973, 1977, 1983) but also adults under hypnosis with a past-life regression therapist (Wambach, 1978). Another technique that seems to elicit reincarnational memory recall is holotropic breathing, developed by the psychiatrist Stan Grof (1998). The explanation for this vast amount of data is quantum nonlocality, and it readily accommodates our reincarnational model of recycling quantum monads. (For details, see Goswami, 2001.)

The subtler aspect of reincarnational data are the phenomena of geniuses and psychological disorders such as phobias that cannot be explained as merely due to the suppressed trauma of this life. There are also a few other phenomena. (See below.) Let's first discuss why people of genius and those with phobias are such compelling proof for the idea of reincarnation.

THE PHENOMENON OF GENIUS

The materialists' explanation of genius and talent in general is genetic. The assumption is that people have "talent genes." It all started with the work of a 19th-century scientist named Francis Galton (1869) called *Hereditary Genius*. Galton claimed, "I propose to show that a man's natural abilities are derived by inheritance, under exactly the same limitations as are the form and physical features of the whole organic world." Galton even provided an impressive list of the genealogy of talented people; for example, 40 percent of the 56 poets that he studied had "eminently gifted relations."

Galton did his work before anybody knew how heredity worked. When genes were discovered, initially there was great excitement over Galton's work, which eventually died down when more data were amassed and with them more understanding of genes. Unfortunately, the fact is that nobody, then or since, has found any talent or creativity genes. Also, we now know that genes do not usually express themselves in any one-to-one correspondence with the macroscopic traits of a person. This is especially true of personality traits, to which at least the environmental conditioning of the current life contributes in a major way. Moreover, the glaring fact that a genius's children are rarely at genius level rules out the genetic inheritance of creativity or genius.

The question can be asked, "Are there any personality traits that contribute to the creativity of a genius?" Certainly, traits like self-discipline and divergent thinking (ability to think about a problem in different ways) contribute to the creativity of a genius, but they are no guarantee. The researcher Donald MacKinnon (1962) did a survey of architects in 1962 in which he found that a noncreative group of people shared 39 out of 40 personality traits with the creative group.

If not personality traits, what then? The case histories of geniuses show that what contributes the most is a strong sense of purposiveness and a psychological drive to explore meaning, especially those archetypal contexts of meaning. It is this drive that Easterners call *sattva*—the mental propensity for fundamental creativity. The psychologist Carl Jung (1971) identified it in people of modern times as an unconscious psychological drive that pulls up archetypal images from the collective unconscious.

Down-to-earth creatives are inventors, who use their minds for situational creativity that Easterners call *rajas*. And let's not forget that the vast majority of people engage neither in *rajas* nor in *sattva*, but engage their minds simply with conditioning responses—a tendency that Easterners call *tamas*. If geniuses possess an uncanny amount of *sattva*, so dominant that neither conditioning nor common situational creativity can obstruct it, where does such *sattva* come from? I believe that the predominance of *sattva* in geniuses can be understood only within a reincarnational framework (Goswami, 1999). One must work out the tendencies of *tamas* and *rajas* and cultivate a lot of godly archetypal qualities before *sattva* dominates the personality. In other words, a genius is an old soul.

Stevenson, the dominant researcher of reincarnational memory, believes on the basis of his data that there is a relationship between genius and reincarnation. Why was Mozart able to play piano so well at the age of three? Why was the Indian mathematician Ramanujan able to add infinite mathematical series at a young age? Genes cannot be a satisfactory answer, from what we have learned about traits. Environmental conditioning? How much special conditioning can you instill in a three-year-old? Ramanujan did not even have any formal training in mathematics until age 10. In addition, consider the odds that, while others of their age were engaged in all kinds of *tamas* and *rajas* activities, these children are taken by such acts of *sattva* as music and mathematics. Stevenson offers many cases in which past-life learned propensity is the only answer to the question of the origins of genius.

PHOBIAS

Phobias, in psychoanalytic terms, consist of avoidance conditioning from childhood traumatic experiences. But Stevenson found many cases where there were phobias but no childhood traumas. Therefore, Stevenson (1974, 1987) attributed those cases of phobias to the reincarnational category. Many more cases have now been reported by past-life regression therapists (Woolger, 1988). What makes the latter work persuasive is that past-life regression therapy is effective in curing the phobias acquired from past lives.

OTHER RELATED PHENOMENA

I will now mention some other cases of inheritance of past-life character. I started this chapter quoting the Dalai Lama. The religious post of Dalai Lama is neither inherited, as in monarchies, nor elected, as in democracies. How do the Tibetans choose their current Dalai Lama? They believe that lamas and rinpoches (*rinpoche* is a Tibetan Buddhist honorific title) are special spiritually complete personalities (quantum monads) that reincarnate in their culture in an ongoing way. Tibetans depend on reincarnational memory recall and even more on reincarnational character traits such as *sattva* dominance, ability to read and recite scriptures, and so on to find the current incarnates of lamas and rinpoches. In fact, the current Dalai Lama was found on the basis of such tests.

The cases of the reincarnational transmigration of vital body propensities are very convincing. I will tell you about an impressive case from Stevenson's large repertoire. The subject in this case was an East Indian man who remembered that in a past life he was a British army officer who served in World War I and was killed by a bullet through his throat. Stevenson was able to verify many details of this man's past life story by visiting the Scottish town of his previous life—details that the subject had no way of knowing in his current life. What makes this case interesting is that this man had a pair of birthmarks on his throat that exactly resembled the bullet wound of his previous incarnation.

An explanation can be given as follows: The bullet that killed this man's physical body was an acute trauma to his correlated vital energies at the throat, especially the vital energy correlated with the skin there. When his physical body died, the vital body trauma translated as a propensity that gave rise to the birthmarks when he was reincarnated in a new physical body.

Xenoglossy, a paranormal phenomenon in which children (and sometimes adults) are able to speak a language not learned in the current life, without any accent (or with the accent of the past), also falls in the category of transmigration of vital body propensity. Our capacity for speaking a foreign language with the correct accent is severely impaired in adulthood because the pronunciation of the vowel sounds demands certain flexibility in using the lips, tongue, and so forth that can be developed only in childhood. If an adult speaks a foreign language with proper accent that she has not learned in childhood of the present life, it can mean only one thing: she inherited the appropriate morpho-genetic field from a previous incarnation.

A spectacular example of xenoglossy and also of channeling (see below) is found in the case of "Lydia Johnson," a 37-year-old housewife in Philadelphia studied by reincarnation researchers Sylvia Cranston and Carey Williams (1984). Lydia was initially the subject of her hus-band's experiments with hypnosis. But soon, with the help of another hypnotist, she began to channel an entity named Jensen Jacoby, a man who lived in a tiny village of Sweden in the 17th century. She pro-nounced the name as "Yensen Yahkobi" and spoke Swedish freely while channeling him. Most tellingly, she would take on his character and rec-ognize only 17th-century Swedish objects and also forget how to use modern tools, such as a pair of pliers.

EVIDENCE OF THE EXISTENCE OF
OUR SUBTLE BODIES FROM SURVIVAL DATA

Quantum monads do not necessarily take rebirth immediately. The evi-dence of their existence between incarnations is called *survival data,* for obvious reasons. These data also provide us direct proof of the exis-tence of our subtle bodies, and in fact, of the quantum monad.

Channeling is the phenomenon in which a person receives commu-nication from a discarnate being. Popular pictures notwithstanding, one can make a theory of channeling using the idea of quantum monad in a straightforward fashion. Realize that, although a quantum monad (soul) cannot collapse quantum possibilities in any ongoing manner when it lacks a physical body, it certainly could do this if it could temporarily borrow a living physical body under mutual (nonlocal) agreement. This is what happens in channeling. A channeler, through mutual nonlocal intentions, becomes correlated with a discarnate quantum monad. Henceforth, for mutually agreed-upon periods, the quantum monad can use the channeler's body to have a physical presence. The proof of this model would be verifying that the character of channelers undergoes drastic change during the periods of channeling.

The phenomenon of channeling has a long, checkered history, but it has finally become scientifically investigable. The idea is to compare specific and measurable neurophysiological performances of the channelers in their normal state and in their channeling state. I will mention three such investigations.

Parapsychologists Gilda Moura and Norman Don (1996) did a comparison of brain wave data of channelers in both states. There is a famous channeled entity called Dr. Fritz that several channelers in Brazil independently have been able to channel. While channeling Dr. Fritz, the channelers perform amazing feats of surgery with fairly primitive instruments. What can be better proof, when the channelers have no previous training in surgery, than this feat? Moura and Don provided even more convincing proof. They measured the brain waves of the channelers in many of their normal states of waking consciousness. Usually the channelers showed beta waves with frequencies ranging around 20 to 30 Hz and *never* exceeding 40 Hz. But while they were performing surgery, their brain waves suddenly jumped in frequency beyond 40 Hz; this showed extreme concentration that they were not even capable of normally. The data proves beyond a doubt that the channelers were using an unusual borrowed propensity, but from where? The only explanation that makes sense is that they were channeling the discarnate quantum monad of a surgeon who had acquired the characteristic of intense concentration.

The channeler JZ Knight was similarly studied, using eight different psychophysiological indicators. The investigators found a marked difference in the observed range for all the indicators between her channeling mode and her normal mode (Wickramsekhara *et al.*, 1997).

More recently, a channeler in Brazil named João Teixeira de Faria, called John of God, has become quite famous for his many documented cases of healing through the energies of love that he channels (Cumming and Leffler, 2006). João was never trained as a medical doctor, let alone as a surgeon; yet, when he channels, he performs skillful surgeries demonstrating a remarkable shift in character. Even his manners, stance, and speech change during the channeling episodes.

I will tell you about one of these remarkable surgeries. The medium João had a stroke that paralyzed one side of his body. Amazingly, however, during this period, whenever João channeled an entity and became John of God, the paralysis disappeared. What about this change of psychophysiological indicators? Even more amazingly, João was able to channel one of the entities of love energy to operate on and heal himself. (He continues in good health.) For further details, read Cumming and Leffler, 2006.

ANGELS AND SPIRIT GUIDES

There are a lot of anecdotal reports of people being guided in their personal lives by angels or what are called *spirit guides*. The famous poets William Wordsworth and Rabindranath Tagore talked about their muses or spirit guides. Is this just the metaphorical expression of exuberant creative experiences, or should such statements be taken literally?

I think that we should take them literally, since the present theory of quantum monads permits it. The idea is that a quantum monad goes through many incarnated lives, learning from experiences in every incarnation until it has become individuated, liberated from the birth-death-rebirth cycle. What then? The quantum monad has no further need to reincarnate as a mental being, but it could be nonlocally available for channeling to whoever can correlate with it.

KARMA AND DHARMA

The complete theory of reincarnation must also contend with reincarnational content memory and certain cause-effect entanglements that may happen between two disparate incarnations. Reincarnational memory is easy to understand; we assume that there is a nonlocal window that is always open between reincarnations. Ordinarily we are not aware of it, but at the moment of death, when ego attachment is extremely tenuous, a person may be open enough to this nonlocal window to have a panoramic view of himself or herself through various lifetimes. Similarly, at the moment of birth, since the ego attachment has not formed yet, the opening of the nonlocal window may allow a panoramic experience of incarnations to be stored in the memory of the newborn.

There are data in support of this theory. In near-death experiences, many people explicitly describe their panoramic vision of this life and sometimes even their past lives. Similarly, while no one remembers offhand the memories of early childhood, especially the moment of birth, such memories have been elicited through special techniques and are found to be consistent with a past-life panoramic vision. For example, Grof uses the holotropic breathing technique for regression to birth or even pre-birth. Many of his subjects recall reincarnational data. I have already mentioned past-life regression therapists. They, too, find it suitable to regress their subjects to very early childhood to extract past-life memories.

How does a cause-effect entanglement take place between two people that continues beyond this incarnation to the next? If two people are correlated through quantum nonlocality and one of them collapses and experiences an event, it becomes a certainty that the correlated partner will experience the collapse, except that the exact time does not have to be specified. It can happen any time in the future, even if the future happens in the next incarnation. So is the nature of quantum nonlocality.

In this way, a cause in this life can nonlocally propagate to the next life to precipitate a nonlocal effect. Theosophists use the Sanskrit word *karma* to denote such nonlocal cause-effect connections between incarnations.

But, of course, there are these mental and vital propensities that are also effects that we carry from one life to another life. I have called these propensities karma in my earlier work (Goswami, 2001), although the Sanskrit word *samskaras* is also used for this particular transmigration.

There is a third connotation that can be given to the word karma. This is the repertoire of supramental contexts that we learn to live, those we carry with us from one birth to another. That is what the Theosophists call our *higher mind*. This is also something that propagates from one life to another and therefore can be called *karma*.

So defined in this way, karma denotes everything that transmigrates from one life to subsequent lives. We then can bring accumulated karma through many lives into the current one. Then, of course, there is the future karma, the karma we collect in our current life.

However, the Eastern literature on reincarnation contains one more concept related to karma—ambient karma (*prarabdha* in Sanskrit), karma that we bring to bear in this life. *Prarabdha* is the portion of the past karma that is responsible for the present body. The idea is that we don't bring all of our accumulated karmic propensities into the current life we are living—only a select number of them.

Surprise of surprises, this idea has been verified by empirical data through the research of a past-life therapist named David Cliness. He has studied many subjects who have recalled various past lives. Curiously, he found that people don't bring all their previously learned contexts and propensities from their past lives to the present one. It is as if one plays poker with them and chooses five out of the deck of the available 52.

We can theorize. Why do we bring a specific choice of ambient karma? Is it because we want to concentrate on a particular learning agenda for this life? This learning agenda is denoted by another Sanskrit word, *dharma* (spelled with a lower-case d to distinguish from Dharma with a capital D, which denotes the Whole, Tao).

This idea of a learning agenda for life may remind you of the wonderful film *Groundhog Day*, in which the hero reincarnates (sort of) from one life to another with a single learning agenda, which is a biggie—love.

One more thing I can say about dharma. When we fulfill the learning agenda that we bring to the current life, life becomes full of bliss. And if we find bliss in our lives, we can conclude that we must be following our dharma. The mythologist Joseph Campbell used to say, "Follow your bliss." He knew.

REINCARNATION AND ETHICS

I earlier introduced the idea of ethics, idealistic ethics. But why should we follow ethics, idealistic or not, if we are behaviorally conditioned beings? In today's social environment, following ethical guidelines often means a personal sacrifice. And ethics are not like physical laws; there is no punishment if you don't follow them. If you don't adhere to the law of gravity and try to fly, you will fail and fall, a reminder that the law is compulsory. Do we similarly fall when we violate an ethical law?

When reincarnation is taken into account in our science, we can see that ethical laws are as compulsory as physical ones. Through our unethical actions, we set up a nonlocal karmic causes that will propagate its karmic effect, its revenge, in a future life. There is no free lunch as far as karma is concerned.

Part Four

Downward
Causation Revisited

In fall 1976, I was motivated. The question "Why do I live this way?" by then had stayed with me long enough to become a burning uncertainty. Not only was I trying to change my research field at work, but also I was doing a lot of meditation in order to change my lifestyle.

There are many ways to meditate, of course, but the one practice that suited me most at the time goes by the name of japa, a Sanskrit word meaning "repeated recitation." You take a mantra, a monosyllabic one preferably, and recite it in your mind over and over. I was particularly interested in one claim that I found in Hindu literature. It is said that if you persist in japa in all your waking hours as best as you can, even while doing other chores, the japa becomes internally established and it continues unconsciously all the time. This stage is called ajapa japa—japa without japa.

By November, my tenacity paid off: in a stretch of seven days, I found myself so settled in japa that I thought it was continuing all the time. I mean, whenever I looked internally, I found that the japa was there. "Well," I thought, "Isn't that interesting? But what happened next was a bigger surprise."

On the seventh day, a sunny morning, I was sitting quietly in my accus-tomed chair in my office doing japa. After about an hour, responding to an urge to walk, I went outside. I remember deliberately continuing my mantra while I stepped out of my office, descended the stairs, and went out of the building and across the street onto a grassy meadow. And then the universe opened up to me. For a split second, I was one with the grass, the trees, the sky, the entire universe. The sensation of connectivity was intensified beyond belief. Concomitantly, I felt a love that engulfed everything in my consciousness—until I lost comprehension of the process. This was what the yogis call Ananda, *spiritual bliss, I knew. The cosmic expansion of my awareness stayed only for a moment or two. A little later, William Wordsworth's words came to mind:*

There was a time when meadow, grove, and stream
The earth, and every common sight,
To me did seem
Apparelled in celestial light
The glory and the freshness of a dream.

("Intimations of Immortality," Complete Poetical Works, *edited by Thomas Hutchinson, revised by Ernest De Selincourt, 1961, p. 460)*

I felt elated for a long time; the bliss of that experience continued unabated for two days before it began to fade. Afterward questions arose: did I have a Samadhi, *the Sanskrit term for the state of pure awareness, which later I was to call* the quantum self experience? *I consulted Patanjali's* Yoga Sutra. *The description of* Sananda Samadhi *(Samadhi with* Ananda) *seemed to fit my experience.*

After a few years, I still recalled the experience with awe and felt that it had inspired me to continue my search. But I also knew that the experi-ence had not transformed me—I remained much the same, except maybe a little more interested in creativity.

Chapter 16

What Does ESP Prove?

Many professionals and experts who advocate a more human experiential dimension to science see clearly how the materialist straitjacket of the current paradigm of science is limiting the possibilities we are all entitled to live. What they don't agree on is the best way to convince the general public and eventually the scientific community about the limits of today's establishment materialist science.

Many of these people believe that it is the research of our paranormal experiences, extrasensory perception (ESP), telepathy, etc., that will lead the way to a new scientific paradigm. Naturally, materialist scientists fight back tooth and nail trying to discredit the paranormal research. And thus arise all the controversies that surround paranormal research today. Is ESP real or is it all a carefully concocted fraud by clever magicians? Paranormal research itself is bogged down by these controversies.

What has gotten lost in all this debate is that we don't need ESP to prove the inadequacy of the current scientific paradigm. As I have amply demonstrated in this book, ESP is not needed to prove the existence of God and downward causation either. Now we can relax and ask objectively: is there evidence of ESP? And what does it mean?

EXTRASENSORY PERCEPTION

Whereas the role of nonlocal quantum consciousness (God) is somewhat implicit in ordinary perception (you don't see the role without much analysis, as discussed earlier), in ESP phenomena nonlocality is explicit; the only analysis we need is to demonstrate the role of consciousness.

Let's set the context by describing a typical distant viewing experiment pioneered by the physicists Russell Targ and Harold Puthoff (1974) and replicated many times by other researchers. One subject looks at a double-blind selected scene or an object; at a distance another subject in a controlled laboratory setting draws a picture or gives an oral description of what his or her partner is observing. What is viewed and the description received of it at a distance are then compared. The experimenter looks for a matching rate that substantially beats the odds of random matching.

Targ and Puthoff made history with such an experiment in their pioneering paper, successfully demonstrating nonlocal transfer of information and meaning from one mind to another. Subsequent experiments verified the efficacy of distant viewing in a variety of ways. I cite here some of the notable ones.

The effect persists even when matching is done objectively via the use of computers (Jahn, 1982).

One of the most stringent protocols used is the *ganzfeld experiment*. A *ganzfeld* (German for "whole field") is created through sensory isolation of the receiver. The receiver is put into a soundproof room. His or her visual field is made uniform and featureless by covering his or her eyes with halves of ping-pong balls and bathing them with uniform red light. White noise is fed into his or her ears through earphones. Many *ganzfeld* experiments have been conducted with good success rates (Schlitz and Honorton, 1992; Bem and Honorton, 1994).

Distant viewing works with both psychic and non-psychic subjects, trained or untrained. (For details see Targ and Katra, 1998.)

Distant viewing works even at international distances (Schlitz and Gruber, 1980).

Distant viewing works between humans and dogs and even between humans and parrots (Sheldrake, 1999).

TELEPATHIC DREAMS

The research of Montague Ullman, Stanley Krippner, and Alan Vaughan (1973) carried out at Maimonides Hospital in Brooklyn, New York, for over a decade, as mentioned in chapter 14, has established the validity of dream telepathy. In their many carefully controlled experiments, a subject (the receiver) was asked to sleep and dream; the dream state was monitored through observation of rapid eye movements and with electroencephalography (EEG). At the sign of REM, the researchers would alert a second subject (the sender) to view attentively a certain selected painting. At the end of each dream period, the receiver would be awakened and asked to talk about his or her dreams. The descriptions of the dreams (see Ullman, Krippner, and Vaughan, 1973, for details) leave no doubt that the sender had affected the content of the receiver's dream through the telepathic transfer of information and meaning.

WHY PARAPSYCHOLOGY IS CONTROVERSIAL

If mind-to-mind nonlocal transfer of information and meaning is so well demonstrated, then why is ESP still controversial? Partly it is because ESP is such an affront to the belief system of the typical materialist scientist that it causes cognitive dissonance. Partly, and more important, it is because one can never guarantee 100 percent replicability of the data. This is actually quite consistent with quantum behavior. But our classical mindset gets agitated whenever the effort to replicate a parapsychological experiment shows ambiguous results.

In this connection, I will now discuss distant prayer-healing experiments. Can you be healed if I pray for you in your name at a distance without even knowing you?

This idea of "other healing" through prayer at a distance was proposed by the physician Larry Dossey in the early 1980s. This hypothesis was duly verified by the double-blind experiment of the physician Randolph Byrd (1988), working with a sample of 393 patients recovering from cardiac surgery. A Christian prayer group did the praying by randomly choosing names from a list of the patients, so neither the physician nor the patients knew who was the subject of their prayers. The healing rate of patients who were prayed for was found to be, statistically speaking, significantly higher than the healing rate of the control group.

But then, in the first decade of the 21st century, researchers convinced the Templeton Foundation to come up with a large grant for doing a repeat experiment on a larger scale. The experiment was carried out by Harvard physician Jeffrey Dusek and collaborators on 1,800 post-coronary bypass patients. The results (Benson *et al.*, 2006) came out negative: no significant healing for patients who were prayed for.

The later experiment was supposedly more carefully planned and carried out, so what should we make out of the results? Were the earlier data faulty because of faulty procedure? One has to be very cautious here!

First of all, if quantum nonlocality is responsible for the distant healing, then all other quirkiness of quantum physics must be allowed to enter the scene. And one quirky part of quantum physics is the statistical nature of quantum events. This statistical nature precludes complete replicability in any case.

Second, as I have pointed out elsewhere (Goswami, 2004), "other healing" ultimately is self-healing, which is creative. So all the uncertainties of creative phenomena further complicate the results. Creativity en masse is very difficult to attain!

Third, there is some evidence that the potency of parapsychological evidence of a particular kind seems to decrease as more studies accumulate and as expectations grow.

Fourth, there is the well-known *observer effect* that Marilyn Schlitz has demonstrated over the years. The intention of the observer affects the result of parapsychological experiments.

Fifth, this is related to previous ones. I personally think that quantum creative downward causation, nonlocal or not, can be used for large enough samples to be statistically significant only to a limited extent. As I pointed out earlier, for large samples, consciousness tends to give up the creative potency of individual choice and allows the probabilistic law of quantum physics to take over. There is a tug-of-war here between the power of intention and the power of randomness to prevail. Quantum physics says that for large enough samples, randomness is bound to prevail.

Under these circumstances, the most sensible research strategy is to choose a sample size and do the experiment tongue-in-cheek. If we get a negative result, the conclusion is clear: the sample size is too large. So we reduce the sample size until we get a positive result. So what does the positive result prove? Statistics unambiguously tells us the odds against the particular deviation from randomness that we have observed in our data.

So you see, by this criterion, both the San Francisco and the Harvard experiments are explained.

IS QUANTUM NONLOCALITY THE CORRECT THEORY OF ESP?

So finally, is the nonlocality exhibited in distant viewing an example of quantum nonlocality? Parapsychologists hesitate to accept this idea because of the *Eberhard theorem*, which purports to have proven that no information can be transferred using quantum nonlocality. I have repeatedly pointed out (Goswami, 2000, 2001, 2002, 2004) that for information transfers between brains and minds, in which consciousness collapses the synchronistic events that constitute the transfer of information, Eberhard's theorem does not apply. And of course, the proof of the pudding is in the eating. My theoretical idea has been verified by the replicated transferred potential experiments (Grinberg-Zylberbaum *et al.*, 1994; Fenwick and Fenwick *et al.*, 1998; Standish *et al.*, 2004).

Let's discuss the latest experiment of this series (Standish *et al.*, 2004), which was designed much like a distant viewing experiment, except that EEG machines were used to demonstrate a "physical" and objective transferred potential. Two subjects were chosen satisfying the following criteria: knowing each other well, having previous emotional and psychological connections, and having experience in meditation and other introspective techniques. One person (the sender) was instructed to send an image or a thought and the other (the receiver) was instructed to remain open to receive any image or thought from the sender during the experiment. The sender and receiver were put into sensory-isolated rooms ten meters apart and their brains were connected to individual EEG machines. The sender was alternately subjected to visual stimulation (stimulus on) and then no visual stimulation (stimulus off). The receiver didn't receive any light stimulation. In spite of this, the EEG of the receiver detected a signal whenever the sender's brain was stimulated (was under stimulus on condition).

As I have stated earlier, the only explanation of the transferred potential is that consciousness collapses the similar events in correlated brains. In this kind of experiment, information *is* transferred nonlocally between brains by virtue of quantum consciousness. By comparing a transferred potential with the very weak brain potential you get for a control subject, you can tell that a subject is sending information and when. Clearly, Eberhard's theorem is violated when consciousness is involved in the transfer of information.

The explanation of quantum nonlocality via quantum consciousness should hold for mental telepathy as well as for distant viewing: consciousness collapses similar events of meaning in correlated minds.

Unfortunately, the parapsychological community seems to be a little squeamish about accepting the primacy of consciousness. Perhaps now that so much evidence has accumulated in favor of a nonlocal quantum consciousness explanation, parapsychologists will see the light (in a flash) and abandon their somewhat hidden materialist prejudice.

Chapter 17

God and the Ego: Co-Creators of Our Creative Experiences

God is the unconditioned agent of consciousness, the collapser with total freedom of choice that gives us true creativity. In our creativity we experience ourselves as the unconditioned quantum self, as the child of God. The ego is the product of psychosocial and genetic conditioning. Our ordinary states of being, both awake and asleep, are dominated by the ego conditioning. One aspect of the divide between materialist science and spirituality is how the two camps view the dual concepts of God and the ego. Although strictly speaking, behaviorists do not formally acknowledge the existence of the ego, since all scientists are hidden dualists, they all surreptitiously believe in a modernist can-do ego. They take themselves very seriously.

In contrast, spiritual traditions are always emphasizing God and the true (quantum) self and undermining the ego. This makes many scientists cry out in support of a not entirely unjustified humanism: "Only the human being is real, there is nothing beyond my humanity!"

A closer look at the creative process resolves this humanistic concern. Which is my true being: ego-me or God, or at least child of God? In the course of our discussion, we will collect further scientific proof for the existence of God.

OUTER AND INNER CREATIVITY

Outer creativity refers to creativity used in the service of the outer world, to create a product in the public arena that anybody can enjoy. Examples are the creative arts, music, and even our creative scientific endeavors. In contrast, *inner* creativity is directed inward to realize the nature of the self. The goal is an inner experience that nobody else can share or necessarily get any benefit from. There is not necessarily any outer accomplishment in inner creativity.

This always creates confusion in the Western culture, which has traditionally been focused on accomplishments. Inner creatives are suspect in this culture from the start. Hence recent practitioners of inner creativity have begun declaring their enlightenment—a confusing idea to say the least. In contrast, the Eastern sage is unequivocal about being enlightened: "The one who says doesn't know; the one who knows does not say." Who is right?

THE CREATIVE PROCESS

I have mentioned, in chapters 6 and 13, the four stages in the creative process as first codified by Graham Wallas (1926):

Preparation
Incubation (unconscious processing)
Insight
Manifestation

The preparation stage is the most obvious: I research what is available as possible answers to the creative question. I create a practice to incorporate the knowledge of others into my being, etc., etc. Obviously my ego is the player here. But from here on, as far as inner creativity is concerned, it is confusing.

The incubation stage is a stage when I relax, "sitting quietly doing nothing." What does doing nothing accomplish? Well, it is confusing to our doing-oriented mind. And yet real creatives know its necessity. I have heard from physicist Hans-Peter Dürr, who was a student of Werner Heisenberg, that Heisenberg always asked his students to wait two weeks after an initial effort and discussion before working on the problem again.

Then there comes the insight. And most creativity researchers agree that insight is discontinuous, sudden. It is not the result of any algorithmic, reason-based thinking. Indeed, many creatives declare after their insight that "God's grace has dawned upon me." In inner creativity, they are even more emphatic about God. Sometimes they declare, "I am God." Creativity researchers find that people always report this experience as a surprise; hence the term "aha! experience" for it. Furthermore, the experience is reported with an uncommon assurance of certainty. "I know. Period. I know on my own authority." Very confusing.

The manifestation stage literally is the embodiment of the insight from the third stage, bringing the insight into form. But this stage also becomes very confusing for inner creativity. In outer creativity, there is a tangible product for everybody to see. The creative gives his or her insight a form that other people can appreciate; people can like it or dislike it, but there is no confusion there. In inner creativity, there is no form to manifest! Is there a form to God? So some masters say, "Before enlightenment, I chopped wood and fetched water; after enlightenment, I chop wood and fetch water." Well?

All confusion disappears upon a quantum reconstruction of the three confusing stages and events of creativity. Let's take them one by one.

INCUBATION

Incubation is just sitting back without working on the problem, as a bird sits on its egg waiting for it to hatch. But what good does sitting do?

Think about processing meaning as processing waves of possibility. As shown in figure 1-2 (page 21), waves of possibility expand to become bigger and bigger conglomerates of possibility in between events of collapse. So, when we are not collapsing particular meanings that we experience as thoughts, the meaning possibilities expand into bigger configurations. Relaxing and sitting quietly without engaging our do-do-do mind increases the gap between thoughts, the gap between collapsed events. This gives the meaning possibilities a greater opportunity to grow into the conglomerate of meaning with a greater chance of containing that particular thought that is the solution to the problem at hand.

If this is still confusing, let me restate it. In between thoughts, in between collapsed events, aren't we unconscious? Obviously subject-object awareness requires a collapse, so who is processing when we are unconscious? But you have to remember that "unconscious" is Freud's term and he did not have the viewpoint of quantum physics. If he had, I am sure he would have used the word "unaware" to denote the unconscious. Because consciousness is always present and, when we are unaware, consciousness processes the expanding waves of meaning possibilities in between collapsed events. In other words, incubation is unconscious processing.

To their credit, creativity researchers knew this even before a quantum theory of thoughts and meaning processing arose. And there is now a lot of experimental data that verifies the idea (see chapter 6).

So unconscious processing has indeed been experimentally verified. When a creative is good at doing it, he or she can process a vast number of possible solutions to a problem, even totally unexpected ones. And wham-o! When the person unconsciously sees the solution, he or she chooses it and collapses the wave. And there is the aha! thought, the insight.

HOW WE GET INSIGHT: DIRECT EVIDENCE OF GOD IN CREATIVITY

Alas! It is not as simple as the description above indicates. If it were, we could go to bed every night, get a good night's sleep, and wake up with an enlightening insight. But what happens? My experience is that before going to bed I am a certain Amit with certain problems, and that when I wake up the next morning, I am still the same person with the same problems, sans enlightenment and new insights.

This has to be surprising because deep sleep no doubt leads to unconscious processing; there is no subject-object awareness during deep sleep. The surprise is abated when we realize that ordinarily we maintain such tight control on what we think that even in our unconscious processing, we don't permit meaning-possibilities that will upset our ego-control. How do we give up control, and what happens when we do that?

Jesus said, "Seek and ye shall find; and when you find, you will be troubled." (This is from the Gospel of Thomas, written by a school of early Christians who claimed the Apostle Thomas as their founder and discovered in a manuscript in Egypt in 1945.) Jesus is giving us a hint. We have to invite trouble to our doorsteps in order to topple the comfortable homeostasis of the ego! We have to generate heat to burn down the house in which the ego resides. We have to make our creative question a burning question!

Once the ego-supremacy of our meaning processing is toppled, room is made for God and the quantum self to enter the scene. Mind you, in our do-do-do consciousness, the ego-control cannot be relinquished; doing involves logistics that involve past learning, the vehicle of the ego. But in your being consciousness, when you are in be-be-be state, just relaxing in between thoughts, God will be processing your meaning possibilities.

I have heard that a radio evangelist often used an interesting metaphor. He would say that in our ego we act like the chairman of a meeting holding a gavel to maintain control. Then he would say with some passion in his voice, "Give that gavel to the Holy Spirit; give up control." Well, he had a point there. The mystic-sage Ramana Maharshi advised his devotees in the same manner when he would say, "Why are you holding on to your luggage? You are on a train."

We cannot give the gavel to God in our do-mode, but we can in our be-mode. If God does the unconscious processing instead of our consciousness in its ego-conditioning, God can look at the possibility conglomerates without influence from previous conditioning. If the conglomerate contains the solution, God has a much better chance of seeing it and choosing it.

We still have to prepare to generate new possibilities for the unconscious to process. We have to create ambiguities that propagate possibilities and stay with these ambiguities without quick resolution. These steps need the ego, the do-mode.

But we always supplement the do-mode with a be-mode, and then again with more do-mode to generate more possibilities. We alternate between do and be, a stage that I call do-be-do-be-do. Michelangelo knew about this encounter between God and ego in the creative process, which he immortalized with his painting on the Sistine Chapel ceiling of God and Adam reaching out to each other (figure 17-1).

FIGURE 17-1. The creative representation of quantum consciousness (God) and the ego (Adam), as depicted by Michelangelo.

Then it happens. One day, the possibility spectrum for unconscious processing contains the right combination that solves our problem, be it inner or outer. And God chooses the insight that we perceive as a surprise aha! moment—because we know we didn't do it. The insight is literally God's grace. Actually, it is more. It is God's choice, the result of which is experienced by us in the quantum self; the ego only makes the mental representation.

In times past, people used to say creativity was God's grace to indicate the acausality of the creative aha! event. Then we discovered that creativity is a quantum leap, but the quantum leap is also acausal. Now we see that even the role of God remains; the quantum leap of creativity is indeed God's grace and God's choice, and it is very direct evidence for the existence of God.

Finally, the question, "Why the certainty of the insight?" Think. Where does the insight originate? Where did we go to take the quantum leap in our God-consciousness? We went to the supramental, discovered a new context, and only then did the insight come.

The supramental is the abode of the archetypes, the real stuff, of which thoughts are representations. Sri Aurobindo calls it the land of truth, and the consciousness that touches it is "truth-consciousness." So with a creative insight, we visit the land of truth in our quantum being, momentarily embracing truth-consciousness. Even when we come back to ego-land and start busily making mental representations, a memory of our trip lingers. This is the certainty that we know, although we may not be able to express what we know with utmost accuracy.

Creativity literature is confusing because there is no universality or commonality in what creatives make of their experiences. In outer creativity, many creatives—materialist scientists are a good example—don't pay much attention to the process and claim that they use the so-called scientific method of reasoning and testing. It reminds me of a TV commercial:

A woman is trying to impress a man with something she is doing, but he does not seem to be interested. Then a light bulb goes off in her head. She goes to the bathroom, uses a mouthwash, and triumphantly returns. "I figured it out," she says.

The ego always wants to take credit. "I figured it out." Fortunately, this is a habit of the lesser creatives only; the great creatives among us—the Einsteins, Bachs, and Gausses—never forget to give credit where it belongs—to God.

MANIFESTATION

Manifestation renders form to the idea generated by the insight. The task is to first give mental expression or representation to the supramental truth, and then to give it any other appropriate physical form that we are capable of creating with our ego capacities.

There is still room for confusion in the case of inner creativity, because in these events the creative often realizes that he or she is identical with God, or at least with the quantum self. There is no content to give form to—or so it seems. Let's delve into this experience in some detail to find the source of confusion and resolve it.

WHAT IS ENLIGHTENMENT?

For inner creativity, depending on the tradition, sometimes the creative is working on the nature of the self, sometimes on the nature of God. In both cases, the final realization is the same, but from two sides: self is God (the self's causal potency, choice, comes from God) or God is self (God can be "experienced" only through the self—the quantum self. Recall Jesus' statement, "No one comes to the father except through me." Jesus is talking from the quantum self consciousness.

Put in the mental mundane way above, the realization experience seems to be quite ordinary. Because you are already grasping it conceptually, the cognitive realization seems to be an easy step. But it would be a mistake to think so.

There is always a particular doubt that we may have about all this God stuff, a knot in our way of thinking that prevents us from realizing God. The realization experience resolves the doubt or the knot in our thinking in an about-face of context. And that is always a surprise!

So there is content after all. In an event of enlightenment, we not only realize our identity with God, but also unravel the knot in our thinking that prevented God-realization. Our thinking determines (as a necessary condition) the way we live; if there is a knot in our thinking, there will also be some knots in the way we live. These knots are responsible for the seams, whereas in truth living can be seamless. What the enlightened has to manifest is living in a seamless fashion—no boundaries.

What is confusing to people about enlightenment is that, like outer creatives, enlightened people express their enlightenment in many different ways, and not necessarily always by changing their mode of living to a seamless one. Why? Because the latter is not as easy to realize as we think. And at first glance, it does not seem necessary to put in so much effort.

Suppose you have an enlightenment experience; you realize you are God and decide to do nothing, because "nothing needs to be done." You decide to live the life of the proverbial Zen master and go on "chopping wood and fetching water" as you did before. Nobody has to know about your enlightenment, and there is no problem.

Unfortunately, this is not the usual trend. After an enlightenment experience, there is almost universally an urge to teach and tell people about your enlightenment (especially in the West), which is necessary to establish credentials. But as soon as people have identified you as enlightened, there will be expectations. Your behavior has to show enlightenment if your teaching is to be credible.

The problem is this. You have realized that you and God are the same being. But has your being shifted to God-being? No, because in manifest existence that is impossible. In the do-mode, the past learning that defines your ego is essential and, as you engage it, the ego enters. The spiritual traditions have an adage, "Even the Zen master has to go to the bathroom." And while in the do-mode, which even includes some of the time you teach, if your unpolished behavior persists with all its uncooked emotions, confusion will arise.

I mention this because every time an enlightened being behaves "badly," the whole movement toward spirituality suffers. Materialists can rightfully question the veracity of knowing God if it cannot produce godly doing, godly qualities in your behavior!

So manifestation is as necessary for the inner creative as it is for the outer creative. As the mystic writer Wayne Teasdale (1999) puts it, "Enlightenment is the awakening to our identity as boundless awareness, but it is incomplete unless our compassion, sensitivity, and love are similarly awakened and actualized in our lives and relationships."

How does an inner creative manifest compassion, sensitivity, and love in his or her life? By taking the arduous creative path to the discovery of these godly qualities in their suchness and following them through in manifestation in living. There is no shortcut. The enlightenment experience is the means to enlightened living, not an end in itself. This is why there is another adage, "Spiritual life begins with enlightenment."

Recognizing the importance of both insight (the sudden aha! experience) and the gradual manifestation of transformation also resolves another controversy: Is enlightenment sudden or is it gradual? In the Zen tradition, there is a Rinzai school that believes in sudden enlightenment and a Soto school that believes in gradual enlightenment. The discussion above shows that both the sudden insight and the gradual manifestation stages are part and parcel of the goal: transformation! Anyone who shortcuts the suddenness of the process in favor of gradual practices never finds certainty as to what the practices mean and where they lead. And anyone who undermines the gradual process of "walking the talk" fools himself or herself about being transformed.

IS TOTAL TRANSFORMATION POSSIBLE?
SAVIKALPA SAMADHI AND *NIRVIKALPA SAMADHI*

In the experience of an insight, God is choosing something new for us without the usual sifting through the reflections in the mirror of memory. In this way, there is an immediacy in any experience of insight. This immediacy becomes most obvious when we creatively look at the nature of awareness itself.

The creative process is the same as described above, except now preparation itself consists of meditation on awareness. So we alternate between meditating on awareness and relaxing—relegating the processing to God-consciousness. At some point, we fall into the primary collapse-state of the subject-object split. As you recall, in this state the subject in awareness is the quantum self and the object is awareness, of course.

What one experiences is the oneness of everything, how the subject and object—the field of awareness—arise from an identity, which is consciousness. In the yoga literature (Taimni, 1961), this is called *savikalpa samadhi*. *Samadhi* means "equality of the two poles of subject and object." *Savikalpa* means "with separation." In other words, in this experience we become aware of the dependent co-arising of the universal quantum self (subject) and the world (object), although the self is already split from the world. We do not ever *experience* consciousness undivided from its possibilities. Any experience, by definition, involves this subject-object split. In other words, *savikalpa samadhi* is as deep or high as we can venture in experience. We see clearly that we are the children of God.

Very confusingly to the ordinary mind, the Eastern literature also refers to another kind of *samadhi,* called *nirvikalpa samadhi. Nirvikalpa* means "without split," without subject-object separation. If there is no experience without subject-object split, what does this represent?

To understand this concept, consider deep sleep. In deep sleep, there is no subject-object split and there is no experience. Yet we have no problem accepting that we all sleep. It is an accepted state of consciousness. *Nirvikalpa samadhi* has to be understood as a deeper sleep in which some special unconscious processing takes place, which is cognized only at the moment of waking—much like the autoscopic vision experience of a near-death survivor who later remembers it upon being revived.

What is the special vision that is revealed from *nirvikalpa?* The mystic sage Swami Sivananda (1987) describes it thus:

> There are two kinds of . . . *nirvikalpa samadhi.* In the first the *jnani* [wise person], by resting in *Brahman* [Godhead], sees the whole world within himself as a movement of ideas, as a mode of being or a mode of his own existence, like *Brahman. Brahman* sees the world within Himself as His own imagination, so also does a *jnani*-wise. This is the highest state of realization . . . .
>
> In the second variety the world vanishes from view and the *jnani* rests on pure attributeless *Brahman.*

Clearly the first kind is the ultimate state of unconscious processing, when we as consciousness in suchness, or God, process the entire world of quantum possibilities, including the archetypes. This is the *nirvikalpa samadhi* posited above. It is *not an experience* but *a state of consciousness.*

The second kind of *nirvikalpa* state that Sivananda describes is called *turiya* in the Vedanta literature. (Vedanta is a school of Hindu philosophy focused on understanding the real nature of reality, especially through the creative insights of the *Upanishads*.) In an earlier book (Goswami, 2000), I made a mistake in trying to explaint *turiya* as a *savikalpa samadhi* without the experience of time. But now I understand it differently and agree with Sivananda that *turiya* must also be a *nirvikalpa* state of non-experience, but a deeper one than reached by *nirvikalpa samadhi* of the first kind.

Can there be an (unconscious) state of consciousness deeper than the unconscious processor of the quantum possibilities of the whole universe? You have to consider this through the involution and evolution of consciousness (figure 9-2, page 128). Realize that quantum possibilities originate with involution, with the supramental being the first stage. What was before then? It was consciousness with all possibilities and no limitation imposed. When all possibilities are included, there is no quality and there is nothing to process; this is one reason why Buddhists call this state of consciousness the great Void and Hindus call it *nirguna*, attributeless.

So what does this approach say about transformation? There is a claim in the spiritual literature of India that people of *nirvikalpa* capacity are totally transformed; their identity completely shifts to the quantum self, except when the ego is needed for everyday chores and ego-functions.

Let's use our model to sort out this possibility. For the achiever of the *nirvikalpa samadhi* of the first kind, the unconscious now processes supramental possibilities. This means that making mental representations of the archetypes and integrating them into behavior would now require little effort. In Jung's language, individuation would become easy without much effort. But there is still "somebody" who is being individuated, who is walking his or her insights in real time. A vestige of identity remains.

The situation is drastically different for a person once *turiya*—unconscious processing in the attributeless state or void—is the case. There is no longer any "thing" to manifest; all desires (*vana* in Sanskrit) of manifestation are now burnt away. So this is *nirvana*, to use the language of Buddha.

So is transformation possible? For *savikalpa* creatives, this discussion has shown that transformation (or individuation) is an arduous journey of many quantum leaps and many manifestations of godliness in one's living. The effort required to be fully transformed or individuated in this way boggles the mind! In truth, what is required is total surrender to God, but how can effort take you to surrender?

This reminds me of a story. A chicken and a pig are looking for some breakfast. They see a diner with a big sign saying "Eggs and Sausages." The chicken is enthusiastic, but not the pig, who wryly remarks, "For you it [eggs] is just a contribution. For me it [sausages] means total commitment."

Now suppose you have the capacity to reach a *nirvikalpa* state of type 1 consciousness whenever you desire it. If your desires were tuned to the movements of God-consciousness, this would be quite natural, wouldn't it? In that case, doesn't it make sense to say that all that you do you would be doing after God's unconscious processing, which would guarantee that it is appropriate? And yet, the very fact that one has desires compromises this exalted state of existence, doesn't it?

Science is telling us unambiguously that only people of *turiya* consciousness are fully transformed in every way imaginable. Obviously, the great mystics of the world, from reading the folklore around them, seem to qualify for this *turiya* level of being. But speaking as a scientist, I must reserve my judgment until more data are available.

Chapter 18

Love Is a Many-Splendorous Evidence of God

Human beings have written more about love than about any other subject. But I feel that almost everybody will agree with that line of a popular song, "I really don't know love at all."

In the 1970s and 1980s, there was a United States senator, William Proxmire, who used to ridicule some of the esoteric subjects of research that academics sometimes chose. I remember how he ridiculed one person's research on romantic love. To this senator, love probably was nothing but one of those genetically built-in things that we do. Love, to him, must have been an epiphenomenon; why waste time on an epiphenomenon when there are such real phenomena as educating children and feeding the poor? For those projects, one can always get a grant without being ridiculed. Okay, those things are important, too; but without love, where would they be?

The flipside is that people who think of love as important also do not see why love should be a suitable subject for science. But at least they would agree that love is a signature of God; wherever love is, God is. God is love, some of them would say.

But what is the signature of love? Is love sex, a feeling, a thought, all of this, or none of this? Is love an expression, whispering "I love you" in someone's ear in a romantic voice? Is love having that wonderful warm feeling in the heart? Is love beyond sex, thoughts, even feelings? Or is love even beyond the beyond, so we cannot talk about it?

I think that with this new science we can talk about love. We can prove that love exists in sex and beyond sex, in words and beyond words, in feelings and beyond feelings—and we can find the signatures of love. This is important because they do tell us about indelible signatures of the divine.

LOVE IS AN ARCHETYPE

I have previously said that biological functions are archetypes in the supramental domain of our being. One of these functions is reproduction, a function that, through the intermediary of a vital blueprint, is physically represented in the male and female sexual organs. Then there is another biological function, the distinction of "me and not me," the archetype of self-world distinction. This one is represented in the immune system. The thymus gland, a heart chakra organ, is a representative of the immune system.

In sexual union, two people are one, physically speaking. There is potentially a source of confusion here for the immune system. So the archetypes of the sexual function and the "me/not me" distinction made a deal. Whenever there is sexual union, the immune system is relaxed about its distinction. And vice versa: whenever the immune system relaxes its distinction and includes someone as "me," sexual union becomes a special urge between two people. At the vital level, this is how this plays out. Whenever there is an excess of energy at the sex chakra, the energy goes to the heart chakra. And whenever there is excess energy in the heart chakra, it flows to the sex chakra. This is, of course, romantic love. Check it out! For romantic love, love and sex go together.

But obviously, our sexual system and our immune system can also act independently, and so can their vital counterparts. Sex is not impossible without feeling the pang of romance,, although one has to be careful not to generate excess energy at the second chakra (see figure 11-1, page 147). More importantly, we have many other important relationships that transcend the "me/not me" distinction, but do not involve sex. In this vein, we can talk about love between parent and child, love between friends, love between servant and master. All these kinds of love can also be felt at the heart chakra.

There are a couple more concepts of love. A special kind of love is said to exist for one's own self, self-love. Can one get a heart chakra response for loving oneself? Yes, one can. And it can soothe you on those lonely nights.

There are more esoteric conceptualizations of love. One is the love of God. What does it mean? Not easy to answer, is it? There is also the concept of loving everyone. Universal love. What does that mean?

Obviously, along with the physical and vital, there is always a mental component of love in all such experiences. We have a meaning-giver, the mind. And we cannot help inviting the mind to give meaning to all of our experiences. So the mind gives meaning to our romantic love experiences, our parent-child relationships, our friendships, our master-servant bonds—all of those experiences, even self-love. And they all also have a vital component. But only romantic love has a sexual component as well, for which the brain's neurochemicals play an important role.

What is so special about love of God and universal love? They can be purely mental and often are. Let's make it a matter of definition. We won't call it "love" unless there is a vital energy feeling experience in the heart chakra. So you really love God when the thought of God gives you a throb or a warmth or a tingle in your heart chakra. And the same applies to universal love. It's not mind f-ing—it's just when the thought of humanity, or a human being, or even a sentient being, makes your heart warm!

We seem to be making progress. Whenever there is some feeling in the heart chakra along with thoughts of love, there is love. So, let's ask, do we know what love is?

Operationally, I would say yes: when there's energy in the heart chakra. And this then is called a signature of love. Feeling in the heart is the operational signature of love. No more, no less.

But we still don't know what love is. We know only its operational signature.

You may remember many love experiences with your mother or father as a child. Then, as a grown-up, you meet a person of the opposite sex and you experience once again that unmistakable signature—that feeling in the heart. Do you know how to behave around this person? Not necessarily. Freudians are not entirely wrong in saying that many people behave in a conjugal relationship as if they were expecting motherly love!

Having a love experience in one context leaves us clueless about how to behave when faced with love in another context. Now do you know what they mean when they say love is an archetype? As with any experience, we make mental (and vital) representations of it, but not a direct physical one, no direct physical memory. We lack that capacity. A representation is never the real thing in suchness; a map is never the territory. This is the fundamental problem in defining love and loving.

QUANTUM SIGNATURES OF LOVE

But having a feeling in your heart is not a definitive signature of love. Why? Because we can be fooled, and we can fake it, and we can fool ourselves.

Suppose, for example, that while thinking of God you are also thinking of your mother, as in the spiritual practice of thinking of God as the divine mother. And you feel that heartwarming feeling. Are you sure you have love of God now? No; it is more likely that the warmth in your heart came from the thought about your mother's love.

Spiritual teachers know about self-deception, and they use their knowledge to prescribe five types of practices for developing the love of God:

1. Meditate on God's love as self-love
2. Meditate on God's love as a servant's or a master's love
3. Meditate on God's love as loving your friend
4. Meditate on God's love as loving a parent or a child
5. Meditate on God's love as loving a lover

I will have more to say about these practices later. But do recognize that they are only practices, ones that will eventually take you to a real love of God.

But how will you know that it is the real experience, not a memory? This is where quantum signatures are useful.

Let's put it another way. Your experience of your mother's love occurred a long time ago, when you were young. Childhood memories are hard to retrieve. But try to think of your first romantic love. How did you experience it?

Perhaps you can remember that there was a suddenness about it, a surprise element. It was a revelation. A sudden insight, an "Oh, I love her/him" moment. The thought came suddenly, as an aha! Of course, the feeling in the heart was there, too.

This suddenness is a quantum signature of love. In the language of the quantum consciousness, you took a quantum leap to the supramental and met the archetype of love, and it directly told you (not in words), "I am here. You've found me." It was only for a moment, that certain thought, that certain feeling. But it was there, unmistakably.

Remember that popular song of the 1970s, with the title and refrain, "I think I love you"? All wrong. It should have said, "I know I love you." Archetypal experiences give us certain knowledge. Of course, it takes a quantum leap for our thinking mind to get that, but because of that certainty, it is worth it.

Caution! Once again, don't fall prey to the thinking that "now I know what romantic love is." You don't. In another context, with another person, you will need to take another quantum leap if you really want to know. But don't worry. You don't have to make the effort. Quantum leaps follow you. Quantum leaps happen to you when you're unaware. That is why we say, "falling in love." We don't do anything. We just allow ourselves to fall. We surrender.

Aside from the quantum leap, is there any other quantum signature of love? There is indeed. Let me tell you about a *Star Trek* episode that gives us a hint:

A fellow in the 23rd century has committed a crime. In that advanced civilization, obviously, corporal punishment is not an option.

But the authorities eventually figure out an interesting punishment. It is decided to send the criminal to an isolated planet in the company of many beautiful women. Go figure! How can that be a punishment?

The women are all androids, machine women. Now do you see the point? Machines cannot give you love; they cannot even give you adequate company. Consciousness is needed to know with; it takes another sentient being to know with, which requires the nonlocal connection, the nonlocal interconnectedness.

(Would a materialist be happy having an android for company or for love? A materialist thinks of himself or herself as an android, so why not? Nevertheless, I think even materialists know the difference—but they pretend, oh, how they pretend!)

So this nonlocality is another quantum signature of love. All kinds of nonlocal events take place around lovers, such as events of synchronicity.

One of the most telling quantum signatures of love is tangled hierarchy. You remember the concept—a circular relationship. Cause oscillates between people in a relationship so you can't tell who is in control. It is agreed that this is not as compulsory as the tangled hierarchy in a brain or a living cell. But let me suggest that even a mild dose of circularity can result in the appearance of self-reference. Consciousness collapses actualities as if there were a third self, the self of the relationship. So a couple becomes a functional unit of and for itself. There is you, there is your significant other, and there is the entity called the *couple* that transcends both of you.

In a simple hierarchy, this will never happen. You may think, well, before women's lib, wasn't a simple hierarchy the standard in man-woman relationships at least? Think again. The simple hierarchy of past societies was only a social imposition. Couples in love have always been able to dance to their own tune. They would have no problem in maintaining a social show of simple hierarchy and yet having a completely tangled love.

To summarize, love has all three quantum signatures: discontinuity, nonlocality, and tangled hierarchy. It is because of these quantum signatures that if you have been in love, you can never be a doubter; you can never be an atheist or agnostic. You directly know that your true being is quantum; how else can you be in a relationship with quantum signatures? Now it is only a matter of time for you to stabilize yourself in your quantum being, God-consciousness. It is only a matter of process.

WHERE IS THE NEW EVIDENCE FOR GOD?

Evidence for God in this discussion of love is different from any of the other evidence I have considered so far. From a strict scientific point of view, I do not know of any systematic, empirical study that proves these quantum signatures of love and hence God's authenticity in this regard.

But looking at it in another way, love is the best field of study if you want to verify God by basing your data collection on your own life. As Bob Dylan sang, "You don't need a weatherman to know which way the wind blows." Similarly, you don't have to let scientists do their laboratory experiments and judge what you should believe. You can collect your own evidence and make your own validation.

But you do need a plan; you do need to engage in a process. Fortunately, in India, where consciousness (or God) research has always been taken seriously, there is a tradition called *Bhakti* (meaning devotion or love). This tradition has developed five ways to study love in your own relationships (the five practices noted above), in your own life, for doing your own personal research on love, and also have some fun doing it (I hope).

And by the way, if you keep a notebook for your research, some day it may inspire a laboratory psychologist to write a grant application for a scientific research project on love, trying to prove the existence of downward causation in that phenomenon. Maybe he or she will get a grant despite the likes of William Proxmire.

CREATIVITY IN RELATIONSHIP

I would like to explore as an example one of the five paths here, of loving God as a lover. In actual practice, we don't start with looking at our lover as God. But we end up there.

A famous God-realized savant of the Vaishnava tradition in India, Sri Chaitanya, asked his favorite disciple about the practices of love.

The disciple said, "Love God as yourself." "This is only superficial. Give me something deeper," said Chaitanya. The disciple said, "Love God as your child."

Now, this is a highly regarded path in India. This was the practice of Krishna's adopted mother, Yoshoda. Krishna was a highly precocious child who performed many miracles, so Yoshoda knew he was God-incarnate and she was devoted to him. But at the same time, she had the duty to discipline Krishna. So you see how the relationship naturally became one of tangled hierarchy.

But even so, Chaitanya was not impressed. "This is also superficial," he insisted. The disciple then cited the practice of serving God as servant and the practice of serving God as a friend, but to no avail. "All superficial," said Chaitanya. Finally, it dawned on the disciple. "Love God as your lover." "That's the one. It is sweet," said Chaitanya approvingly.

But the practice is not always sweet. It starts as sweet, romantic love. When the neurochemicals run out, sweetness ends for a while. If we can endure and love returns, it is sweet again. But to regain the sweetness, one needs creativity, a quantum leap.

When the neurochemicals dry up, sex becomes more mechanical, and having sex does not automatically raise the vital energy to the heart. This is when problems begin. The many suppressed negative emotions of a romantic relationship erupt into conscious awareness, giving rise to defensiveness. In disagreements this becomes fuel to precipitate a fight. This is the time to begin the serious practice of loving our mates unconditionally. And this requires the creative process and a committed partner.

How do we begin the creative process toward unconditional love? Like all creative processes, it starts with preparation. Read the many wonderful books available on the subject from psychologists and spiritual teachers. Going together to a mutually-agreed-upon therapist or marriage counselor is not a bad idea either. Practice awareness to recognize the suppressed emotions that erupt in those moments of defensiveness. The two of you can participate in a limited amount of analysis of your fights, but always with the full awareness that analysis will not solve the problem even though momentarily it may seem to do so.

The fundamental aspect of inner creativity that people miss is that creativity requires ambiguity. Without ambiguity, our unconscious processing stage of creativity never breaks free; the processing revolves around the conditioned memories of past events under ego-control. Remember what Jesus said: "Seek and ye shall find; and when you find, you will be troubled" (Gospel of Thomas). He meant that you will find trouble if you want to go deeper with unconscious processing. The trouble, the ambiguity, is automatic in a romantic relationship gone awry. Your intimate relationship is no longer unquestionably with a lover. Sometimes she or he offers you the olive branch; other times it is the dagger (figure 18-1).

With ambiguity, unconscious processing will create a spreading of the possibility waves of meaning that contain new possibilities beyond

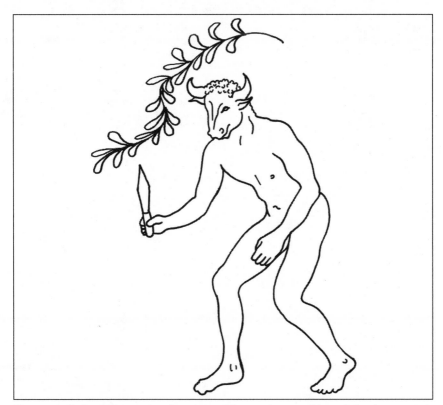

FIGURE 18-1. A minotaur, showing the ambiguous juxtaposition of the dagger and the olive branch, of animality and humanity.

241

past ego conditioning. And now God can enter as a processor of your unconscious; God is always interested in new configurations.

Now you just follow the usual setup of the creative process: intentions, open mind, do-be-do-be-do, and so forth; and then insight. But remember, it may take a lot of little insights to hook the big one. For example, before you can love unconditionally, you go through recognition of your mate as the "other." This is already a quantum leap from your romantic relationship when your mate was just an extension of you.

After the insight that catapults you into unconditional love, you are able to see your lover as God and a possibility of a tangled-hierarchical relationship. This insight you now have to manifest in your behavior.

Once your relationship with your mate is manifestly tangled-hierarchical, you can start a tangled-hierarchical relationship with God as your lover. And, of course, you have to go through the creative process once again, this time with God as your partner in relationship.

Chapter 19

Evidence for Downward Causation in Spontaneous Mind-Body Healing

Of the many examples of mind-body healing, there is a subclass called *spontaneous* healing. This constitutes a spectacular example of a definitive signature of downward causation and, therefore, of the divine.

Spontaneous healing is healing without causal medical intervention. Healing may be triggered by a variety of stimuli, medical procedures, and sometimes just plain intention and faith. In science, unusual phenomena often give us more clues about a particular system. So what is the explanation of this particular unusual phenomenon?

Examples of spontaneous healing, some of them as dramatic as the overnight vanishing of a malignant tumor, are abundant in the literature (Chopra, 1990; Weil, 1995; Moss, 1984; O'Regan, 1987).

What do the data say on the spontaneous remission of cancer? The Institute of Noetic Sciences researcher Brendan O'Regan (1987),

243

who did perhaps the most extensive survey on the subject, categorizes the three kinds of spontaneous remission cases: 1) pure remission, with no allopathic treatment after the diagnosis is made; 2) remission with some treatment after diagnosis, but for which the treatment is clearly unsuccessful; and 3) the most unusual kind of remission, in which the "cures are sudden, complete, and without medical treatment," associated with spiritual cures.

It is the third class of remission cases that offers us the most clear-cut evidence of downward causation—the creative quantum leap.

THE QUANTUM PHYSICS OF QUANTUM HEALING

First, let's explore a little theory by way of explanation. *Mind-body disease* consists of physical ailments in which the imposition of wrong mental meaning sets up disharmony in our vital and physical bodies. So mind-body healing must involve changes in the mind's meaning-context that result in the malfunctioning of the vital and the physical bodies. Sometimes this change in the context of meaning processing by the mind can be brought about simply by reshuffling old contexts. This is when the continuous techniques of mind-body medicine, such as biofeedback and meditation, are effective. But in cases of spontaneous healing, the contextual shift could not have taken place at the level of the mind itself. In those cases, mind-body healing is a misnomer for the healing that takes place.

The most profound contexts of mental thinking come from the supramental domain of consciousness; to change from an old context to a truly new one, we are required to leap to the supramental. This leap is a discontinuous quantum leap, which is why this type of healing is called *quantum healing*.

This term, *quantum healing*, which I have discussed earlier (chapter 13), was creatively intuited, albeit in rudimentary form, by the physician Deepak Chopra (1990). In the 1980s, Chopra was searching for an explanation of spontaneous self-healing. Somebody asked him about the cure for cancer, and he said, "If a patient could promote the healing process from within, that would be *the* cure for cancer."

Earlier, the Christian Science pioneer Mary Baker Eddy had the similar idea that if the mind could discover that all disease is illusion, then healing would follow. In this way, both Chopra and Baker Eddy introduced the idea of healing as self-discovery. But Chopra, being part of the quantum age, was able to go one important step further. He said, "Many cures that share mysterious origins—faith healing, spontaneous remissions, and the effective use of placebos, or 'dummy drugs'—also point toward a quantum leap. Why? Because in all of these instances, the faculty of inner awareness seems to have promoted a drastic jump—a quantum leap—in the healing mechanism."

To see clearly the dynamic role that the quantum leap of insight plays, it may help to further analyze what is involved in these kinds of cancer cures (Weil, 1995). There is perpetual pressure on the cells of the body to become malignant—a condition in which they do not die at the expected time, do not stay in the same place, and in general do not conform to the regular cellular laws of behavior. But malignant cells are not cancer, only the seeds of cancer, and they distinguish themselves by displaying abnormal antigens ("not me") on their surface membranes. A normally functioning immune system, whose job is to distinguish between "me" and "not me," can recognize and destroy these malignant cells. Cancer takes hold only when for some reason this normal immune system function is inadequate, due to a physical or a vital body defect, or suppressed, for example, through an energy block at the heart chakra via excessive intellectualism.

For a healing, we have to boldly recognize the healing power of consciousness, of downward causation with freedom to choose. Consciousness has the requisite wisdom (in its supramental compartment) and the mechanism (choosing a new context for the mental processing of emotions). It also has the power to discover what is needed, to make the quantum leap of insight. And it can manifest the insight, by unblocking the vital feeling at the affected chakra, thus unblocking the movements of the associated vital blueprint and also reviving the correlated physical organ with proper organ function. The spontaneous healing of cancer is due to the sudden onset of such a dynamic surge in immune system activity that the cancerous growth is destroyed within days, sometimes even within hours.

Suppose the immune system malfunction is due to the suppression of feelings in the heart chakra, arising from faulty mental processing of love-related meaning. A quantum leap to the supramental is accompanied by a contextual shift of the processing of mental meaning. This frees the blockage of feelings that correspond to our consciously experiencing the movements of the vital blueprint of the immune system, at the heart chakra. This then can have the desired dynamic effect on the immune system, in the form of reactivating its program of hunting down and killing cancerous malignant cells with such vigor as to effect very rapid healing.

EXPLORING THE CREATIVE POWER OF DOWNWARD CAUSATION THROUGH SELF-HEALING

Conservatives in the medical profession sometimes dismiss cases of spontaneous remission of disease, labeling them as the placebo effect. In truth, faith in a doctor's word, as in the placebo effect, gives a patient only a glimpse of his or her own healing capacity. To truly manifest this capacity, one must use the entire program of creativity, going through all the stages of the creative process that culminates in a complete change of the context of one's living.

In this way, as both Mary Baker Eddy and the philosopher Ernest Holmes (1938) indicated, a life-threatening disease not only poses a danger but also provides an opportunity to explore the transformative power of downward causation. Said Holmes, "Healing is not accomplished through will power but by knowing the Truth. This truth is that man is already Perfect, no matter what the appearance may be." Quantum healing is about regaining wholeness; it is transformative.

For the remainder of this chapter, I will closely follow the exposition in my book *The Quantum Doctor*.

If quantum healing really involves creativity of the mind, can we develop a program of action for healing ourselves based on this idea? It is true that creativity is acausal. But it is also true that engaging in the creative process—with its four stages of preparation, incubation, insight, and manifestation—helps bring about creative acts. What would fully engaging in the creative process entail in the case of mind-body healing?

Suppose that, instead of thinking they are getting some sort of medicine that activates the placebo effect, patients operate under the "burning" conviction that they already have the requisites for healing, which they need to discover and manifest through creativity.

The first step of such a creative endeavor is preparation. Patients would research their diseases (with help from their physicians, of course) and meditate on what they find. Such meditation would readily expose the habits of suppression or expression of emotions, as the case may be, that contributed to each disease. Some of the root causes of mental stress accumulation would also become clear. The speed of mental processing—hurrying and rushing—is one. Augmenting the pursuit of desires with accomplishments, anxieties, and daydreaming is another. So the purpose of the preparation stage would be to slow down the mind and to make it open and receptive, especially in its response to feelings.

In the next stage, the patients would try various new (to them) techniques of mind-body medicine. Here, collaboration with their personal physicians would be, a tangled-hierarchical collaboration, of course, that serves the quantum creative process much better than simple hierarchy. This is the stage of creativity in which we use unlearned stimuli to generate uncollapsed possibility waves of the mind and the supramental, but we don't choose among the possibilities. Since only choice can create an event of conscious awareness, what I am referencing is unconscious processing without awareness.

There are well-known cases of "art therapy" in which people are able to heal themselves by submersion into beautiful, spiritual healing art, but this does not work for everybody. How does art therapy work? These people must be visual, capable of visual imagination. For them, the mental imagination of healing inspired by the artwork very soon gives way to unconscious processing that opens up to a new vista of healing possibilities. Sooner or later, a seemingly inconsequential trigger precipitates the quantum leap of insight: simultaneously the new supramental context and its mental gestalt appear manifest in conscious awareness. The insight leads to the corrective contextual shift in how the mind handles emotions. Manifestation of the insight begins at once: freed from the shackles of habitual mentalization, feelings and vital energy movements at the affected chakras become unblocked, leading to healing of the correlated organ, sometimes quite dramatically.

There are some reported successes in treating cancer patients via the use of creative visualization (Simonton *et al.*, 1978), for which the above scenario applies. Here is a particularly poignant description of one person's quantum healing through visualization:

> When I was in Mexico, I had started having pain in my chest. I went across the border and got an MRI scan, which showed a mass on my thymus connecting to the aorta. I decided just to wait, but a scan six months later showed it was still there.
>
> I decided to spend a week at Carl Simonton's healing center in California, and I imaged "sharks eating cancer cells" as they recommended. But toward the end of the week, I had this extremely vivid, spontaneous vision that wasn't on the program. I saw a mass on my thymus as a piece of ice that just started to melt in these big, amazing drops. I've never in my life had this kind of clear image just come up by itself. And I knew instantly the drops are just teardrops. My whole life, through all the losses, I'd never been able to cry. Now there was this melting away of the oppression I'd been feeling; the deaths and the abuse in my childhood, the unresolved relationship with my ex-husband. The emotion was suddenly available, and it felt so powerful.
>
> Four months later, I had another MRI, and the mass was gone—there was no sign of it. I had no new treatment. Whatever this mass had been, they said the only way they could tell it had ever been there was from the previous two tests. (Quoted in Barasch, 1993, pp. 273-274)

Clearly, the experience released the depression of emotions accumulated through a lifetime. And there is no doubt that the experience was sudden and unexpected, a genuine quantum leap.

A spontaneous remission, in this way of looking at things, is the result of a creative insight, of our ability to choose "the healing path" out of the myriad possibilities generated by unconscious processing. This choosing is the work of downward causation of quantum consciousness—God.

How does one experience this choosing of healing insight, the associated quantum self experience? Experiences vary. The example quoted above was a vision.

The physician Richard Moss (1981, 1984) talks of a cancer patient who attended one of his workshops. During the workshop, she was tired and defiant and was not responding to the various attempts by Moss to energize her. But at some point Moss broke through her shell and she responded by spontaneously participating in a group dance. This led her to a tremendous aha! experience. The following morning, the patient woke up feeling so good that Moss felt compelled to send her for a checkup. Miracle of miracles—tests showed that her cancer was gone.

The patient in Moss's anecdote experienced the more usual "aha!" of creative insight. But patients also report the experience of making the choice itself, the moment when the purity of the healing intention is crystallized. As an example, here is the physician Deepak Chopra's (Chopra, 1990) account of the healing of a cancer patient through sudden insight:

> ... A quiet woman in her fifties [,] came to me about ten years ago complaining of severe abdominal pains and jaundice. Believing that she was suffering from gallstones, I had her admitted for immediate surgery, but when she was opened up, it was found that she had a large malignant tumor that had spread to her liver, with scattered pockets of cancer throughout her abdominal cavity.

> Judging the case inoperable, her surgeons closed the incision without taking further action. Because the woman's daughter pleaded with me not to tell her mother the truth, I informed my patient that the gallstones had been successfully removed. I rationalized that her family would break the news to her in time ....

Eight months later I was astonished to see the same woman back in my office. She had returned for a routine physical exam, which revealed no jaundice, no pain, and no detectable sign of cancer. Only after another year passed did she confess anything unusual to me. She said, "Doctor, I was so sure I had cancer two years ago that when it turned out to be just gall-stones, I told myself I would never be sick another day in my life." Her cancer never returned.

This woman used no technique; she got well, it appears, through her deep-seated resolve, and that was good enough. This case ... I must call a quantum event, because of the fundamental transformation that went deeper than organs, tissues, cells, or even DNA, directly to the source of the body's existence in time and space. (Chopra, 1990, pp. 102-103)

The final stage of the creative process—manifestation—is also important. Manifestation is not complete with only the reactivation of the glands that are needed for the normal functioning of the organ(s) involved. After the remission takes place, the patient has to bring to manifestation some of the lifestyle changes that are commensurate with the shift of mental context and the processing of feelings, if the remission is to be stable and permanent. For example, a lifestyle that produces excessive intellectualism and defensive reactions must give way to a more balanced one of integrated head and heart.

Let's discuss the case of the former *Saturday Review* editor Norman Cousins and his self-healing from a condition called ankylosing spondylitis, a degenerative disease that causes the connective tissue in the spine to wither away. According to experts, Cousins' chance of recovery was only one in 500. In desperation, the patient stopped standard medication and substituted mega doses of vitamin C, all in full consultation with his physician. But most important, the patient decided to submerge himself in happiness; he watched funny movies (for example, W.C. Fields and the Marx Brothers) and read his favorite comic books for a time. And miraculously, Cousins completely recovered from his condition and resumed his very productive life.

I am convinced that Cousins went from a serious disease to healing more or less by following the stages of the creative process. The first stage, his hobnobbing with standard medicine and getting its concept of the disease, was preparation. The second stage, watching funny movies and reading comic books, allowed him the all-important relaxation of the "being" mode of creativity alternating with the "doing" mode of taking vitamin C ("do-be-do-be-do"). Eventually he got his quantum leap, which led to recovery. And from all accounts, he did make lifestyle changes—manifestation of his insight.

There is a lot of similarity between what I am advocating here and what Christian Scientists already practice. However, there is one important difference. In the strict application of Christian Science, no medical intervention is permitted. There is nothing in creative quantum healing to suggest that we cannot simultaneously apply the techniques of conventional and alternative medicine. Sometimes, as in the case of cancer, this may be necessary to keep the physical body alive to allow time for the creative quantum leap to take place. It is reported that even Norman Cousins, in the case cited above, used homeopathy while he precipitated his quantum healing.

Part Five

Quantum Activism

*I*n 1999, I had the unique opportunity to join a group of *scientists at a conference in Dharamsala, India, whose specific purpose was for the scientists to dialogue with the Dalai Lama about applying the new paradigm ideas in science and integrating science and spirituality into our social systems. You can get a feel for the flavor of the conference from the documentary* Dalai Lama Renaissance. *What actually happened at this conference was very educational to me.*

In short, the 30 or so scientists—and most of us knew one another—became very competitive as to who among us should have the best opportunity to present his or her ideas to the venerable Dalai Lama. The battle became so vicious that we settled for the ridiculous compromise that each person would get to present his or her work in two minutes.

I still distinctly remember my embarrassing attempt to summarize many of the ideas that you read in this book in two minutes. Of course, it didn't work, and the Dalai Lama's face showed that he was quite unimpressed. The same fate frustrated almost every one of us; the Dalai Lama showed animation only twice—when a psychologist discussed education and when somebody raised the political issue of the future of Tibet. No wonder somebody summed up our frustration with the comment, "Your Holiness, we come to you as hungry ghosts. ..."

The point is that when we grow up in a materialist society, as all the assembled scientists did, there is no way to escape early materialist conditioning. The competitiveness we all displayed (including me) was deeply rooted in us. Materialism, by denying the importance of meaning, makes one vulnerable to negative emotions.

All our social institutions have fallen prey to negative emotions. The cause of this can largely be traced to the prevalent materialist worldview of the last six decades. How to change it? We can make a beginning with quantum activism—using the transformative power of quantum physics to change ourselves and society.

I will end this preamble with one more anecdote about our meeting with the Dalai Lama. When someone complained to the Dalai Lama about the bitter infighting among the scientists, the Dalai Lama laughed and laughed and simply said, "That is to be expected." This not only helped to dissipate the bitterness among us, but also proved to me that the Dalai Lama is a highly transformed person.

Chapter 20

Quantum Activism: An Introduction

Q uantum activism begins when we change our worldview from a matter-based one to one based on quantum physics and the primacy of consciousness. We have begun right thinking and we ask: now that we know how to think properly about our world, what should we do about it? We have taken the first step toward becoming quantum activists.

Quantum physics, when interpreted through the philosophy of monistic idealism, is transformative. Right thinking—giving up myopic materialist ideas and embracing God, downward causation, and the importance of the subtle bodies in our lives—is the first step of a transformative journey. There is more.

Spiritual traditions regard the journey of transformation as spiritual—a journey toward the spirit, the unmanifested reality, leaving the manifest world behind. The transformative journey of a quantum activist is different.

Our science within consciousness is telling us that the manifest material world is designed to represent the possibilities of the unmanifest better and better as time proceeds through evolution.

Transformation is important primarily to serve the evolutionary play of consciousness and only secondarily for personal salvation in spirit.

So, as quantum activists, we do not leave the world; instead we live in the world with the right attitude. We combine right *thinking* with right *living*.

RIGHT LIVING

How can we live so as to serve the evolution of consciousness in manifestation? It turns out that this is a balancing act.

Materialists see life as biased toward the material end, hardly leaving any space for meaning, let alone the supramental. The conventional people of spirit live life weighted heavily toward the spiritual. The path of the quantum activist is the middle path: the subtle and the spirit are valued, but so is matter that makes representations.

To the materialist, life is the playing out of genetically, evolutionarily, and environmentally conditioned programs. Only the ego exists. Spending life in the service of the ego is the goal. One becomes equivocal about meaning and values. To the spiritual seeker, the object is to become embodied spirit—the quantum self. The goal is to perpetually live in the quantum self of (inner) creativity. One becomes confused about the manifest world.

The quantum activist lives in growing balance between the two extremes. The activist knows that it is as important to manifest the *content* (the insight) of a quantum-self experience as to manifest the *context*. And the manifestation of the content requires sophisticated structures of the mind, many repertoires of representing meaning. For the quantum activist, living in the ego and living in the quantum self have to be balanced in a life focused on personal growth.

It is now customary to classify mental health into three categories: pathological, normal, and positive. Psychotherapists mostly work with patients who need to be lifted from pathological to normal. *Normal* mental health here is defined as a state in which a person is capable of normal ego-sustaining activities and of maintaining relationships, and is fairly balanced emotionally. *Positive* mental health is enjoyed by those who are happy much of the time, who are creative, who are more or less independent of their environments, and who have some capacity for unconditional love, a sense of humor, and some other, less important qualities (see Maslow, 1971). Everybody has the potential to move from normal mental health to positive, which is the essence of personal growth. For quantum activists, it is a prerogative.

We become interested in personal growth when we begin to deal with the great meaning questions that lie dormant in the ego-identity of ordinary mental health. "What is the meaning of my life? What am I doing here?" These kinds of questions. We are no longer satisfied with the status quo of the ego-homeostasis. The inquiry about the meaning of life launches us into self-inquiry, and even further, into an inquiry about the nature of consciousness itself. And when we discover the evolutionary nature of the movement of consciousness in manifestation, we align ourselves with that movement.

BALANCING THE GROSS AND THE SUBTLE

Materialists avowedly live at the gross level, although I suspect that surreptitiously many of them do partake of the subtle—feelings, meanings, intuition, values. Likewise, the spiritual seeker avowedly ignores the gross, but secretly may appreciate the color of money quite a bit—survival instinct, you know. For the quantum activist, there is no conflict of worldview there. Both gross and subtle are necessary for making manifestation possible; both are important. The quantum activist directs his or her attention to both.

The quantum activist considers the nuances of the material dimensions of life, such as making a living, but does not get lost, does not identify with his or her professional persona as such. The quantum activist openly enjoys and explores the subtle—feelings, meanings, and values—soul food.

In the past, and to some extent even now, spiritual pursuits have been identified with practices like meditation, prayer, the reading of good books, and even celibacy. If love is included as a spiritual practice, it is in the form of agape or compassion—objective love. This aspect of spirituality is important for the quantum activist also, but it is not everything. The quantum activist engages in spirituality in everyday living as well. Not only does the quantum activist explore love as charity and service to others, but also in intimate relationships, even carnal relationships. In this way, the spirituality of quantum activism is close to the tradition of *tantra*.

It behooves a quantum activist to be aware of the difference between pleasure and happiness. Pleasure eventually separates. There is nothing wrong with temporary separation from wholeness, of course, provided we practice moderation. Happiness, however, is always the result of wholeness. You cannot go astray with happiness.

BALANCING THE VARIOUS SUBTLE DOMAINS

Of utmost importance to the quantum activist is the balancing of the various subtle domains of the self—feeling, thinking, and intuition.

To the materialist, thinking is everything; rationality is supreme. Even the fact that the progress of scientific research itself depends on quantum leaps of intuition does not influence the strict rationalism of the materialist.

Mystics are one step ahead of the materialists; they embrace both the rational and the intuitive planes. But they do not put them on an equal footing. Invariably, spiritual traditions tend to denigrate creativity involved with processing meaning in outer manifestation—creativity in the arts, humanities, and sciences. This will not suffice for quantum activists, who must balance both outer and inner creativity in their lives. Both are important for our evolution.

Mystics also, by and large, tend to avoid base feelings and negative emotions, never bothering to transform them. This has led to much misconception about the behavioral usefulness of the so-called mystical "enlightenment." What good is enlightenment if it does not enable a person to behave with equanimity even when faced with provocations that call for anger, greed, or lust?

The inner creativity of using the mind has been the traditional popular tool of spiritual seekers with the objective of attaining insight or *samadhi* (enlightenment, *satori, gnosis,* or whatever else you call it). Love, which additionally requires working with and transforming (vital) energies of emotions, is left out in this male-oriented tradition, so much so that in the 1980s the spiritual women of America rose in protest to coin phrases like "feminine spirituality" and "the feminine face of God." We quantum activists have to integrate this dichotomy and practice creativity with love.

The greatest challenge I envision for a quantum activist is the transformation of negative emotions into positive ones, that is, achieving emotional intelligence (Goleman, 1994; Krishnamurthy, 2008). Look around you. In all our social organizations, negative emotions are rampant. If we quantum activists don't know how to transform them, how can we effectively ask others to show restraint and emotional maturity?

The transformation of negative emotions involves, in addition to mental creativity, creativity in the domain of the vital. The challenge is to engage the creative process simultaneously in the domains of the mind and of the vital energies. The practice of unconditional love discussed in chapter 18 falls in this category. Quantum healing at the vital level (see Goswami, 2004) gives us opportunities to invoke vital creativity.

BALANCING THE STATES OF CONSCIOUSNESS

Materialists emphasize only the waking state of consciousness, for obvious reasons. A staunch materialist would rather stay awake and make money all the time or pursue other material ventures if he or she could. This may account for the popularity of stimulants in our culture, in spite of the habit-forming destruction they bring.

Spiritual seekers in general seek states of *samadhi*. As such, they don't pay much attention to their dreams. And they endure the ordinary waking state and deep sleep only because they have no choice.

But this won't suffice for the quantum activist. Remember: he or she is interested not only in *samadhi*, but also in the insight gained in creative experiences, including *samadhi*, to change the quality of waking and dream life. To the quantum activist, creative insight and *samadhi*—visiting the supramental intuitive plane—are important, but so are the waking and the dreaming states. To pay attention to the dream state, you must engage in dream analysis; it has much to contribute to your spirituality.

Not only waking awareness, but also dreams can be used for personal growth and for spiritual transformation. In this connection, you should pay great attention to archetypal dreams. These dreams are about the laws of movement for all the bodies contained in the supramental body. Of these, the laws of the material body are quantitative. But the laws become progressively less quantitative and more thematic as they move to the vital, to the mental, and on to the supramental. For example, mental meaning revolves around certain contexts such as love, beauty, truth, and justice—all qualities. These are the great Platonic archetypes.

The origin of Jungian archetypes in dreams can now be understood. Because of necessities of manifestation, limitations of expression occur and certain themes are suppressed from our normal waking awareness. The confluence of these suppressed themes is what makes up the *collective* unconscious, whose suppressed themes are universal; this is in contrast to the Freudian *personal* unconscious, whose suppressed themes are personal. In dreams, once again, our normal guard against suppressed themes is weak, thus raising the possibility of their surfacing. Indeed, they surface as the now well-known symbols of Jungian archetypes: the great mother, the hero, the shadow, the trickster, the anima and the animus, and so forth. Working with these archetypal dreams makes the unconscious conscious once again, as we come to terms with the learning agenda of our supramental body, what the psychologist James Hillman (1992) calls the soul's intent. (See also chapter 15.) We become open to creativity and the other characteristics of positive mental health.

An example is in order. Consider the archetypes of the anima and the animus. The anima is the archetypal woman in man—the possibility waves in male minds that correspond to the female, but are suppressed because the male's genetic and environmental conditioning may find their expression inappropriate. The animus similarly is the suppressed male possibilities in women. Why should we change these conditioned tendencies for suppression? Because the anima in men also represents the quality of receptivity, an essential quality for creativity, so men need to integrate the anima. Similarly, women need to integrate their animus because it enhances willpower, which is necessary for the three P's of the creative process—preparation, perseverance, and production.

In the mid-1980s, I was struggling with spirituality and rekindling my creativity for some time, but to no avail. The conventional spiritual practices were not doing me much good.

One night I dreamed I was looking for water in a stream, but the stream seemed to be dry. Then I heard a voice: "There is no water there. Look behind you." I did, and it was raining. So I soon found myself walking in the rain along the meadows, and a very pretty young lady joined me. The walk became quite joyous among all that water, and I also had a joyous conversation with my companion.

When the meadows came to an end, there was a house, and it seemed that my young companion was about to enter the house.

"When will I see you again?" I asked.

"I am going to London. When I get back, I will be waiting for your call," she said, and disappeared. I returned to walking joyfully through the meadows.

This dream, which is clearly a classic anima dream, was crucial to my development. It encouraged me subsequently to focus my energy on anima integration, which was the key I was missing.

Let me mention one more thing. Dream analysis can be much easier with the help of another person or persons. In other words, when we are working on our personal growth through dreams, who is to take the place of the psychotherapist? One answer is to find a spiritual teacher who works with dreams. (Many do.) During the period 1987-1989, I worked extensively on my dreams with the spiritual teacher Joel Morwood. An easier and more appropriate avenue is to join a dream group or to create one. As mentioned before, I did that too.

REINCARNATION: FIND AND FOLLOW YOUR BLISS

Materialists don't believe in reincarnation; there is no room for reincarnation in materialist science. Conventional spirituality allows for reincarnation, but the emphasis is always to go beyond the birth-death-rebirth cycle. For the quantum activist, things are quite different.

I discussed the East Indian concept of dharma in chapter 15. This concerns the archetypal learning agenda that we bring with us into each life; we choose our karmic propensities accordingly. When we fulfill our dharma in this life, we experience bliss.

Quantum activists are in no hurry for liberation from the birth-death-rebirth cycle. Accordingly, we must each pay attention to our inherent karmic propensities and use them to fulfill the learning agenda, dharma, of this life. Following our bliss in this way frees us to serve the evolutionary movements of consciousness.

EVOLUTIONARY ETHICS

I introduced the subject of ethics in two previous chapters in connection with the soul and reincarnation. There is still another way to look at ethics—evolution.

There is one remaining problem with all three ethical philosophies discussed in the previous chapters: why do some people follow ethics and others don't? At least in times past, maybe fear of hell or desire of heaven was an incentive. But few take heaven and hell seriously enough anymore to sacrifice selfishness. I submit that the reason so many of us, even today, try to be "good" in our daily living, with so much unethical behavior in our societies, is evolution. There is an evolutionary pressure that we experience as a calling, and we respond.

It follows that ethics need not be looked upon as religious or spiritual, nor is there any need to compromise and adopt the scientific (materialist) ethics of "the greatest good for the greatest number" or subscribe to a bioethics driven by the genes. We can solidly base ethics on the very scientific notion of evolution.

Let's define an evolutionary ethics. As I have discussed elsewhere (Goswami, 1993), a good ethical principle that seems inescapable for us in the idealist sense is this: *ethical actions must maximize the creativity of people, including ours*. Evolutionary ethics goes one step further: *ethical actions must maximize the evolutionary potential of every human being*.

As an example, let's consider a serious ethical problem. You and a group of scientist colleagues have discovered the technology for developing a new weapon of mass destruction. The ethical question is whether to develop the weapon or not. In a previous age, the excuse that others would sooner or later develop the same weapon to use against you would have made you prone to violate ethics and develop the weapon, even though there was no immediate threat. Patriotism creates an ambiguity. This is exactly what happened with the atomic bomb. But evolutionary ethics is not like the religious ethics of the previous age. It advocates the same ethics for all of humanity, an objective ethics needed for the evolutionary future of humankind. So you don't need to be equivocal, and you would be able to reject outright the thought of developing the new weapon.

RIGHT RELATIONSHIP WITH THE ENVIRONMENT

Hopi Indians are known for their emphasis on "right relationship," not only with people and animals, but also with their environment at large, extending even to the whole planet.

In the inward journey of conventional spirituality, right relationship with the environment is often ignored. No doubt this has led to the modern movement of *deep ecology*. Once we become established in an evolutionarily ethical relationship with all human beings, it is time to ponder our ethical responsibility to all creatures, great and small, including our nonliving environment. In short, let's ask, What is our responsibility to the planet earth, to Gaia?

Deep ecology (Devall and Sessions, 1985) not only requires abiding by a few rules for preserving our ecosystem and passing a few laws to reduce or prevent environmental pollution, but also means taking actions in ambiguous situations that demand a creative quantum leap.

When you take such a quantum leap, you realize one astounding fact: *I choose, therefore I am, and my world is*. The world is not separate from you. When we do this en masse, we will leap into a truly Gaia consciousness, which has arisen in human vision from a different context (Lovelock, 1982).

RIGHT ACTION

So finally, what is the plan of action of the quantum activist? Using the oft-quoted phrase of the Hindus, what is the karma yoga of the quantum activist? For Hindus, karma yoga is applying spiritual practice in the middle of real life by doing selfless service. This is an important practice of many spiritual traditions, especially of Christianity and Sotto Zen. For a quantum activist, karma yoga is extended toward selflessly serving society and the world with evolution in mind.

In our current materialist culture, accomplishment is the standard-bearer. When one acts with an accomplishment orientation, any action, even one that is seemingly selfless, always tends to strengthen the ego—the accomplisher. To undermine the accomplisher within us, we must not take ourselves too seriously. In other words, we dance, but always lightly, not caring how anyone thinks of us, not even how we think of ourselves.

RIGHT LIVELIHOOD: BRINGING MEANING BACK INTO OUR SOCIETY

Three of our recent great social accomplishments—capitalism, democracy, and liberal education—all originated from the idea of making meaning available for everyone to process. But now, under the aegis of materialism, the pursuit of meaning has degenerated into a pursuit of power. This is a major deterrent to our future evolution.

One central purpose of quantum activism is to bring meaning back into our social institutions, on which we depend for our livelihood. So whatever your real-life situation in our society, it will provide you with ample opportunities to practice karma yoga for your quantum activism, with the goal of shifting from the pursuit of power back to the pursuit of meaning. For example, if you are a businessperson, clearly business is your arena for quantum activism, where you can follow your bliss and where you can restore meaning in your life.

Chapter 21

Summing Up

In the 18th century the emperor Napoleon summoned the scientist Pierre Simon, Marquis de Laplace, and asked him why he had not included God in his latest book on celestial movement. To this inquiry, Laplace is supposed to have replied, "Your Majesty, I haven't needed that particular hypothesis."

It has been a long time since Laplace's era, but even today establishment science's "proof" against God's existence consists of the insistent disclaimer, "We don't need that particular hypothesis."

If the establishment science's crusade against God is directed at the dualistic God of popular Christianity, a mighty emperor on a throne in outer space doling out rewards and punishment, I am sympathetic to their point of view. But when it seems to include a dismissal of all causal agencies outside of the material world, then it is time for all good people to take note and reject this "old" science.

This book shows that all the sciences—physics, biology, psychology, and medicine—need the hypothesis of downward causation, introduced as conscious choice from quantum potentia, to make sense of their most basic principles and data. The agency of this downward causation, quantum consciousness, is what the esoteric spiritual traditions of the world call God, popular views notwithstanding.

The theory and facts presented in this book as scientific evidence of the existence of God speak for themselves. Consider:

We cannot find a better physics than quantum physics. Its theory is sound; its verification data is flawless.

We cannot find a better interpretation of quantum physics than the idealist consciousness-based interpretation, simply because it is the only one that's paradox-free.

We cannot find a better metaphysics on which to base our science than the primacy of consciousness, simply because only this philosophy is inclusive of all our experiences, "everything that is the case." (This quote is taken from the *Tractatus Logico-Philosophicus* by Ludwig Wittgenstein, which starts, "The world is everything that is the case.")

We cannot understand creativity without the idea of quantum leaps of discontinuity.

We cannot explain of the fossil gaps of evolution without the idea of downward causation and biological creativity.

We cannot find ways to distinguish between life and nonlife and between conscious and unconscious without the idea of tangled hierarchy.

We cannot resolve the paradoxes of the subject-object split in our normal perception without the concepts of downward causation, tangled hierarchy, and nonlocality.

We cannot understand the vast amounts of experimental data of our interconnectedness without the nonlocality of our consciousness.

We cannot understand the vast amounts of data of near-death experiences and reincarnation without the idea oo the nonphysical subtle bodies.

We cannot understand acupuncture and homeopathy without the concept of nonphysical vital energies.

We cannot understand meaning, how the body suffers with its distortion, and how it contracts disease without the concept of a nonphysical mind.

We cannot understand why physical laws exist, why altruism exists, why ethics and values influence our conscience, and how healing works without the concept of a nonphysical supramental body.

We cannot have a proper science of ethics without the hypotheses of downward causation and subtle bodies.

We cannot understand spontaneous healing without the concepts of downward causation, quantum leaps, and subtle bodies.

We cannot understand ourselves without knowing God—our deepest causal being, our quantum consciousness.

We cannot know our evolutionary future and prepare for it without appreciating the evolution of consciousness.

God exists. Realize It. Live It. Love It. Evolve the energies of love.

To paraphrase the poet Rabindranath Tagore:

In the violent night
Under the thrust of death
When humans break through
Their earthbound conditioned limits,
Will not God's unlimited heavenly glory,
Supramental intelligence,
Show itself?

It will. Our dark night of the soul, the materialist interlude, is almost over. In this dark night, we have done our creative processing that Indians call *tapasya* (spiritual practice that burns out impurities) and we are developing a new science to guide us in our evolution toward the supramental. There is still some way to go, some time to wait; the night is not over yet. But the early light of the new dawn is visible for those who can see.

Epilogue 1

Approaching God and Spirituality Through Science:
An Appeal to Young Scientists

I have heard that a young scientist once approached the mystic Jiddu Krishnamurti and asked, "How can I do science and still be spiritual?" To this Krishnamurti replied, "You can be spiritual by doing science to the best of your ability." But this was another era (the 1970s and early '80s), when the integration of science and spirituality was practically unthinkable. Now that such an integration is not only thinkable but also demonstrable, Krishnamurti's answer misses the mark. In this epilogue I will answer this question for young scientists: you have an opportunity to realize God-consciousness and arrive at transformation while doing science, if you approach it in the right way, with right thinking.

But this answer needs elaboration, lots of it. The following is an elaboration in the form of an imaginary dialogue.

Such a dialogue serves another purpose. There is an old saying that old scientists never change their mind on a paradigm shift, but they do

269

die. The paradigm shift that is presented in this book will not convince any die-hard old-timer. But by being a little more technical than the main body of the book, this dialogue may help provide added incentive to young scientists to approach science differently. The young scientists are the key to the paradigm shift and its exploration. (Nonscientists can skim over the in-depth science to get to a more general discussion later.)

Young Scientist: I appreciate what you have presented in this book, but I have so many questions, and I see so much ... incompleteness in your arguments.

Author (smiling): And I thought I had been very thorough. Give me an example.

YS: Well, the most glaring oversight is neglecting to mention that there are many other solutions to the quantum measurement problem besides the one you discuss, and no more radical. There is the many worlds theory, a favorite with many physicists. The transactional interpretation—another favorite. You could have at least stated the truth—there are viable alternatives to bringing consciousness into physics.

A: There may be, but I haven't seen any yet. The two alternatives you mention are dualistic. They are assuming that the final apparatus for measurement is *nonmaterial, without saying so*. They camouflage it well, of course.

YS: I don't understand.

A: Remove your blinders. The many worlds theory looks viable because the authors hold out the attractive promise that discontinuous quantum collapse is not necessary. They theorize that one can be completely faithful to the mathematics of quantum physics and still solve the measurement problem by realizing that a measurement involves a proliferation of parallel universes, each containing the manifestation of one facet of the quantum possibility wave involved. Do you see the camouflage?

YS: Frankly, no.

A: The measurement still involves a measurement apparatus for amplification of the signal, right?

YS: Yes, of course.

A: But all measurement apparatuses, if they are material, become waves, superposition of possibilities, when they interact with a wave of possibility, don't they?

YS: I am not sure.

A: Think about it. This is the crucial point of the camouflage. Comprehend what John von Neumann (1955) was trying to tell us through his celebrated von Neumann theorem.

YS: Remind me.

A: Now I have to be a little technical. All interactions in quantum physics must keep the basic linear structure of quantum physics intact; they must all conserve probabilities. In technical language, all interactions must amount to unitary transformations.

YS: I suppose I have to agree.

A: But in the many worlds theory, the interaction with the measurement apparatus is doing something more than a unitary transformation: it is splitting up the universe into branches. The same is true for the transactional interpretation, where it is assumed that the interaction with the measurement apparatus somehow triggers the emission of a possibility wave going backward in time. In this way, these models are also taking us outside quantum physics. They are proposing measurement apparatuses that are not made of quantum-physics-obeying matter.

YS: How about the variations of the original many worlds theory?

A: The same criticism always applies.

YS: I see.

A: Look, it is the same difficulty that Niels Bohr encountered in his Copenhagen interpretation, except that he did not use any camouflage, so most physicists saw the difficulty right away. Bohr also said that the measurement apparatus is different, that it obeys classical Newtonian deterministic physics and so it does not become a wave of possibility. And this no physicist would agree with, even before von Neumann established his theorem.

In fact, if you read carefully, you will see that all the possible alternatives to the conscious observer interpretation become problematic

with the von Neumann theorem. And this includes all efforts to eliminate collapse. I have written about this in some detail (Goswami,
2002, 2003).

YS: How about David Bohm's (Bohm, 1980) interpretation? Is that not
a viable alternative?

A: Unfortunately, no, it is not. Bohm's is a modified quantum physics,
an approximation of quantum physics. There is no reason to sacrifice
the elegance of quantum physics for an approximation that works in
a rather clumsy way, just to keep consciousness outside its parame-
ters. Actually, I have done better. The physicist Mark Cummings and
I were able to show that the Bohmian approximation surreptitiously
assumes collapse anyway. It is too technical to delve into here, but I
have discussed it elsewhere (Goswami, 2002).

YS: All right, you've convinced me. There is no other interpretation of
quantum measurement that does the job properly except the one
you discuss. Shall we move on?

A: What? You're not giving me an opportunity to pitch my main idea,
that quantum measurement problem gives you an enormous opportunity to rediscover God, realize God, within science?

YS: Now you have made me properly curious.

A: OK. In the *Upanishads* of the Hindus, they discuss discursive methods with the same objective of God-realization. One such method
consists of discussing and meditating on the problem of the nature of
happiness and suffering.

YS: It would be interesting to get a glimpse of that.

A: Maybe on another occasion. For you, the quantum measurement
problem is more appropriate, if you approach it with the question:
what is the nature of consciousness that can collapse a quantum possibility wave without introducing any paradox?

YS: (a little excitedly): Yes, yes. I see what you are saying. I liked your
approach to the paradox of Wigner's friend. It was quite enlightening to realize that consciousness has to be nonlocal. I wouldn't say
I took a quantum leap, but it was very satisfying. But tell me this.
Why didn't the satisfaction last longer, and why did skepticism
come back?

A: Satisfaction is a transient phenomenon. It doesn't last. Skepticism is good; it is the indicator that you did not take a quantum leap. But now you say you understand that consciousness has to be nonlocal. This is a very good starting point for spiritual work. It is called *faith*.

YS: I see.

A: It is an intuitive glimpse at reality. Now you have an opportunity to dig deeper. Can I experience nonlocality of consciousness in my being? How do I do that? Do I meditate? Do I delve into psychic experiences?

YS: Those questions never arose in me.

A (smiling): Nonlocality is not your thing; it doesn't turn you on. Now take the question of circularity of the observer effect. Here you have another opportunity to go deeper.

YS: Tell me more.

A: You understand that circularity is tangled hierarchy and is self-reference?

YS: I suppose.

A: Go deep. Why is self-reference, the separateness of subject and object, arising? It is because we are stuck at the same level as the object. In a quantum measurement, we identify with the brain, a physical object in space and time. Notice how space is created by the semipermanence of all macroscopic physical objects, semipermanence due to the sluggishness of their possibility waves. Notice how time is created by all those memories of past collapses in the brain. So you look at yourself as a physical object in this world of space and time. The perception is too real to give up.

YS: But I intellectually like the idea that there is an inviolate level, an underlying whole—quantum consciousness, God—that is the cause of the collapse, the origin of downward causation. The example of the liar's sentence is a good one to elucidate the importance of the inviolate level.

A: I am glad that your intellect is tickled. But here is an opportunity to go deeper.

YS: Deeper? How?

A: Consider for a moment a different model of ourselves. Not of quan-

273

tum measurement, but of how our autonomy may arise. I have mentioned holism in the book. Self-reference via quantum measurement is one model of the self of, let's say, a single living cell. But holists have another model. The thesis is that the self arises as an emergent property of self-organization as a "whole" that is greater than the parts and cannot be reduced to the parts. So far so good?

YS: Yes, very good. And the holists would say that this emergent self has autonomy, has free will in the sense that the experience of free will cannot be reduced to the components. Couldn't our free will be like that?

Well, I know what you will say. This emergent free will is ultimately determined, determined from the lowest material level, because there is no causation other than upward causation in the model. But in your model also, our free will is ultimately determined by God's will. What's the difference?

I say the holists' model is better because it satisfies my principle of parsimony. Why introduce God when we don't really need that concept?

A: Don't we? Let's see. By the way, do you know that some holists have delved deep into Buddhism?

YS: What has that got do with anything?

A: In Buddhism, our free will is an appearance. When we look deep, we discover we really are empty of any so-called "self." This fits very well with the holists' theory of the self.

YS: Then what's your point?

A: The point, my friend, is that Buddhism is not nihilism. Emptiness does mean nothingness, but it is no-thing-ness. Emptiness is infinite potentia. It is potential fullness, as Hindus would put it.

YS: I still don't understand. So Hindus and Buddhists disagree about the nature of ultimate reality. What else is new?

A: But they *don't* disagree, don't you see? Emptiness is no-thing, it is full of quantum possibilities, and it is fullness in potentia. When consciousness is empty of the known, the playground of our conditioned ego-self, room is made for the unconditioned to come through. Buddhism does not talk much about the unconditioned, but it is

implicit. They leave it for you to find out as a surprise. The unconditioned is another name (perhaps a very accurate name) for God with
downward causation, the same as in all spiritual traditions.

YS: So why can't the unconditioned be the elementary particles at the
base level, their upward causation, only now we are experiencing
them directly without the interference of past conditioning?

A: If that were so, if the spiritual work of deconditioning ourselves just
led to an unconditioned "will" arising from the unfiltered movement
of the elementary particles, there would be no transformation. Our
behavior would show a haphazard mixture of order and chaos. Isn't
that so?

YS: I suppose. So transformation is your proof of downward causation?

A: Transformation is the most obvious proof, as emphasized in one of
my earlier books (Goswami, 2000). But don't forget the quantum
measurement problem. Holism does not solve the quantum measurement problem either. And if you think about it, it does not really
have enough explanatory power for explaining biological evolution.
Nor can it resolve the neurophysiological dilemma of the subject-
object split in perception.

YS: You mean nobody has been able to demonstrate those things yet!

A (smiling): Well, creative evolution via downward causation and biological creativity is a manifest theory. It is not promissory. For biological creativity, we also need the morphogenetic field and the mind
and the supramental. For resolving the subject-object split in perception, we need to apply quantum measurement to the situation.
Holists are never going to demonstrate that feeling, meaning, and
physical laws or even ethics are due to the holistic emergence from
complex interactions of elementary particles going through many
levels. There is a category difference. But we are moving away from
our subject.

Transformation is important and it is impossible to incorporate into
any materialist theory, holism included. If that is the convincing you
need, start with that.

YS: OK. What's the next step?

A: The next step is to recognize what you are transforming.

YS (startled): Huh? What am I transforming?

A: The internal chaos that exists in all of us, that causes our suffering and our separateness. It is a chaos of meaning and feeling and, occasionally, a chaos of value, is it not?

YS: I would agree with that.

A: Transformation is a transformation of the context we use to process meaning, feeling, and value, right? So looking at what we are transforming, we immediately discover these nonphysical bodies of our consciousness. We now have all the ingredients of a new scientific paradigm: downward causation and the subtle bodies.

YS: And your point is ...?

A: In days past, religions also had the concept of downward causation. They used it as a magic wand, as the cause of all unexplained phenomena, mostly material phenomena. Now the neo-Darwinists have such a magic wand: natural selection looked upon as adaptation. But the downward causation of the quantum God-consciousness is not a magic wand. It is an empowerment, giving us real free will, freedom of choice. When we discover it, we are empowered to change our internal environment first, bring order there. And eventually, even make our external environment better.

YS: So as scientists we should be encouraged to study downward causation involving the subtle bodies, not only because it gives us a new set of phenomena and problems, but also because when we study it, we cannot but empower ourselves to transform. The scientist is no longer separate from the subjects of his or her study.

A: You got it. In this way, the scientist joins the evolutionary movement of consciousness toward the soul level of being.

YS: Thank you. I would like to be a scientist in search of the soul. Thank you indeed.

Epilogue 2

Quantum Physics and the Teachings of Jesus:
An Appeal to Young-at-Heart Christians

I n dedicating this epilogue to you as Christians, I am hoping that you are true to your namesake as disciples of Jesus. You may have some embodied teacher at the moment; you may have had many such teachers in your past; but Jesus has always been your great teacher, what a Hindu would call a *sadguru*, a true guru, a teacher who is stabilized in spirit.

The important question for all Christians is, of course, this: is the God that science is rediscovering the same as the Christian God? I have sometimes reassured you this is the case: the new science God is the same God as that of esoteric Christianity, and that of Christian mystics such as Meister Eckart and St. Teresa of Avila. Nevertheless, I can demonstrate this by directly comparing Jesus' teaching with the lessons of quantum physics, which will remove all doubt. Or so I hope.

Jesus was one of the great spiritual masters of all time. He gave his teachings in terms of puzzles and paradoxes. This is already similar to the lessons of quantum physics, which also creates puzzles and paradoxes in our minds. Both Jesus and quantum physics are talking about reality, but are they talking about reality the same way? This is the great question. If they are talking about reality in terms of identical metaphors, however puzzling and paradoxical these metaphors may be to the rational mind, there is reason to conclude that there is convergence. Fundamentally, they are the same. Jesus' God and the quantum consciousness God are one and the same.

THE BASIC FABRIC OF REALITY

Consider the idea of the basic fabric of reality. Materialists say that reality at its base level is reduced to building blocks called elementary particles, such as quarks and electrons, and that causation is upward from this base.

But quantum physics says otherwise. In quantum physics, there are no manifest material objects independent of subjects—the observers. In quantum physics, objects remain as potentia, waves of possibility, until they are brought into manifestation through the act of observation. Quantum objects are waves of possibility, but possibility of what? They are the possibilities of consciousness. Consciousness, not matter, is the ground of being, in which matter exists only as possibilities. Through the act of quantum measurement or observation, consciousness converts possibility into actuality, by collapsing waves into particles or things, at the same time splitting itself into a subject that sees and objects that are seen.

What does Jesus have to say about the fabric of reality? It is quite unequivocal, albeit a little sarcastic to upholders of material supremacy. (All quotes from the *Gospel According to Thomas* are taken from Guillaumont *et al.*, 1959.)

> If the flesh has come to existence
> because of the spirit
> it is a marvel;
> but if the spirit has come to existence
> because of the body
> it is a marvel of marvels.
> (Thomas, p. 21)

Jesus says, resonating with quantum physics, that flesh has come into existence because of the spirit, not the other way around.

It also pleases me greatly that Jesus said, "Spirit gives life; the flesh counts for nothing" (John 3:6; 6:63). No support here for the materialist theories of the origin of life, including the theory of emergent autopoiesis (self-creation) of the holist. But Jesus' saying resonates fully with the quantum idea that life originates from tangled-hierarchical quantum measurement by consciousness. Of course, being a mystic, Jesus underestimates flesh, matter. In the new science, we now can spell out the contributing role of matter—it is to make manifestation possible and to make representations of the subtle.

NONLOCALITY AND TRANSCENDENCE

Popular Christianity posits God and the Spirit as separate from us; and this dualism, of course, is where most scientists find Christianity unscientific. If God is truly separate from us, then how can we receive God's guidance and love? How can flesh, material substance, interact with the nonmaterial divine?

Quantum physics has a different take on this. God is not separate from us; God is indwelling in us, in our unconscious. Consciousness is the ground of all being, which includes us. This resonates well with Jesus' statement, "My Father and I are one." And if you interpret this statement to mean that Jesus is only talking about himself, that only he is the "son of the Father" and therefore identical to Him, the gospels say otherwise. Jesus repeatedly tells his listeners that all are children of God; they just have to realize it:

If you know yourselves
then you will be known
and you will know that
you are the sons of the Living Father.
(Thomas, p. 3)

Quantum physics also says, "You and I are one." Consciousness or God collapses similar possibility waves in both of our brains if we are "correlated," giving every one of us the opportunity to verify this idea of oneness. This idea has even been verified in the laboratory. If consciousness can collapse waves of possibility in your brain and mine simultaneously when we are correlated, we must be connected through our consciousness, which is nonlocal and a unity for both of us and, by inference, a unity for all of us.

The concept of nonlocality is subtle. It also implies that you and I are connected without any signals through space and time. So our connection through consciousness transcends space and time. Yet we are also manifestations of the same consciousness; it is consciousness that is immanent in us.

What is Jesus' view on these subjects? Let's take his well-known statement: "The kingdom of God is everywhere, but people don't see it." So Jesus certainly knew and preached about God being immanent in the world. But is this an animistic worldview? Let's not be hasty. Here is another famous quote from Jesus:

If those who lead you say to you
"See, the kingdom is in heaven,"
then the birds of heaven will precede you.
If they say to you: "It is in the sea,"
Then the fish will precede you.
The kingdom of God does not come visibly,
Nor will the people say,
"Here it is," or "There it is,"
because the kingdom of God
is within you.
(Luke 117:20-21)

And again, says Jesus:

But the Kingdom is within you
And it is without you.
(Thomas, p. 3)

The kingdom is not localizable; we cannot say it is here or there or at any one place. It is both outside and inside, both transcendent and immanent. All this is resonant with the message of quantum physics.

CIRCULARITY, TANGLED HIERARCHY, AND SELF-REFERENCE

One of the most interesting features of quantum physics is the circularity that exists in the observer effect: there is no collapse without an observer, but there is no (manifest) observer without collapse. The circularity is a tangled hierarchy of logic that gives us self-reference, the subject-object split that the observer experiences. Amazingly, Jesus was already intuiting this when he said;

> If they say to you:
> "From where have you originated?"
> say to them:
> "We have come from the Light,
> Where the Light originated through itself."
> (Thomas, p. 29)

"Light" here refers to the Holy Spirit, the quantum self in the quantum physics language. We have come from the Light: our individuality is the result of conditioning. The Light originated through itself, through circularity, tangled hierarchy.

JESUS AND THE QUANTUM SELF

I previously said that the final stage of spiritual enlightenment is reached when one is steadily situated in quantum God-consciousness whenever one is processing unconsciously. I think that Buddha reached this last stage of enlightenment, because there are plenty of anecdotes about his equanimity.

But Jesus lived for a short time, and much of that time is shrouded in mystery and controversy. The accounts we read do suggest that Jesus sometimes engaged in meditation, but the gospels are more full of what Jesus said and miracle stories.

For Jesus, these miracle stories are very telling. Miracles are, of course, not performed in the unconscious, so they are not suggestive of whether Jesus was steadfast in God-quantum consciousness. But miracles do suggest that on those occasions Jesus acted from the quantum self, or what in Christianity is called Holy Spirit or simply Spirit, and he chose from possibilities beyond all limitations.

The idea is that ordinary creativity—vital and mental—involves the laws and contexts that are codified in the supramental domain of consciousness. Miracles that are in violation of physical laws, such as Jesus' conversion of water into wine, are suggestive of creativity that transcends the supramental laws of physics. In other words, the person who is doing it has unconscious access to possibilities beyond the supramental, beyond the limitation of the quantum laws of physics, in the bliss body itself, in *turiya* consciousness.

So it is not a surprise that Jesus sometimes spoke from this quantum self or Spirit consciousness, creating much confusion, as in his celebrated statement:

I am the way and the truth and the life.
No one comes to the Father
Except through me.
(John 14:6)

The Christian church has used these words in attacking other religions, other faiths. But it is confusing to Easterners, too, when they compare Jesus' statement with such statements by Eastern sages as "Those who are enlightened do not say, those who say are not [enlightened]." Should not a person whose self-identity has shifted beyond the ego to the Spirit at least be humble? By all accounts Jesus was a very humble man when he was in his ego, acting from an ordinary state of consciousness. Confusions of both groups disappear if we consider that when Jesus makes this kind of statement, he is speaking from the relatively rare nonordinary state of the quantum self. It is the same nonordinary state from which he performed miracles that superseded physical laws.

And if you still have doubts that Jesus did sometimes talk from the nonordinary state of the quantum self, why else would he have made such a statement as this: "Before Abraham was born I am" (John 8:58)? Or for that matter, make a statement like "Learn and understand that the Father is in me, and I am in the Father" (John 10:38)? A person has to be in the tangled-hierarchical state of the quantum self in order to realize the circularity that gives rise to the human condition.

JESUS AND CREATIVITY

The disciples said to Jesus, "Tell us what Heaven's kingdom is like."

He said to them:
It is like a mustard seed—
smaller than all seeds,
but when it falls on the tilted earth
it produces a large tree
and becomes shelter for all the birds of
Heaven. (Thomas, p. 15)

What does this mean to you? Why is Jesus emphasizing a seed that is smaller than all seeds? Could it be that an insight is a glimpse from the supramental, smaller than other seeds—the usual thoughts that clutter our psyche? And yet when this seed falls on the tilted earth, it becomes a large tree on which the birds of heaven take shelter. And yet when an insight comes to a prepared person (tilted earth), it produces a transformed mind (a large tree) where many of the archetypes (birds of heaven) can be represented (can take shelter). Thus Jesus knew of the three stages of inner creativity—preparation, insight, and manifestation. He did not mention the stage of unconscious processing here, but he does mention it elsewhere:

And he said, "The Kingdom of God is
As if a man should scatter seed upon
the ground, and should sleep and rise
night and day, and the seed should
sprout and grow, he knows not how. . . ."
(Mark 4:26b-29)

The phrase "he knows not how" clearly acknowledges that some of the processing in inner creativity, growing the kingdom of heaven within oneself, is unconscious.

Jesus himself attained perfection, and he encouraged people to do the same:

Be perfect, therefore, as your heavenly Father
is perfect. (Matthew, 5:48)

And what does perfection consist of? It is being situated in command of the supramental beyond the mind—the realm of dualities:

Jesus said to them:
When you make the two one,
and when you make the inner as the outer
and the outer as the inner,
and the above as the below,
and when you make
the male and the female into a single one
so that the male will not be male
and the female not the female,
then shall you enter the Kingdom.
(Thomas, p. 17)

Many people are hesitant to endorse the Gospel according to Thomas as being fully authentic. If that is the case, can we trust that these words are authentic, that they did come from Jesus? In my opinion, if these words were inserted by another author, this other person would have to be wise as well. We should look for historical evidence for him or her. And until we find such evidence, we may as well ascribe these words to Jesus.

Do you see how tuned the discoveries and conclusions of the new science are to Jesus' teachings? Jesus said, integrate inner and outer. Usually, mystics emphasize the inner and downgrade the outer world. But not Jesus; he knew that God is both. Just as the gross/outer attracts the materialist, the subtle/inner may seem attractive to connoisseurs of consciousness. But we must resist the temptation and make the outer and the inner as one.

Similarly, we need to integrate the above and the below, the transcendent and the immanent, the wave and the particle in our quantum language. We must avoid the tendency of the religionist to embrace the transcendent in preference to the immanent. Likewise, we must avoid the materialist indulgence of embracing only the immanent while denying the transcendent.

Finally, why does Jesus say to us to integrate male and female? This does not seem to be a concern of quantum physics, does it? But I think Jesus is not talking about integrating our male and female psychological tendencies—Jungian style. I think he is talking about male-yang and female-yin in the sense of Chinese medicine, the creative and conditioned ways we process our subtle bodies. We need to integrate and use both methods always. Creative quantum leaps must be followed up with manifestation. Then shall we transform—then shall we enter the kingdom of heaven.

IF JESUS WAS TRANSFORMED, WHY WAS HE SO UNFORGIVING?

The philosopher Bertrand Russell wrote:

> There is one serious defect to my mind in Christ's moral character, and that is that he believed in hell. I do not myself feel that any person who is really profoundly humane can believe in everlasting punishment . . . . There is, of course, the familiar text about the sin against the Holy Ghost: "Whosoever speaketh against the Holy Ghost it shall not be forgiven him neither in this world, nor in the world to come." . . . I really do not think that a person with a proper degree of kindliness in his nature would have put fears and terrors of that sort into the world. (Quoted in Mason, 1997, p.186).

It is a reasonable expectation that a transformed person would see only God potential in another human being. Indeed, the transformed saint Ramakrishna's disciple Vivekananda said this about his guru: "My guru has the most beautiful eyes, because he cannot see evil in any one anymore, he only sees the divine potential." Indeed it is well documented that Ramakrishna treated prostitutes and Brahmins with the same love, causing much unhappiness among members of the latter group.

285

But if Jesus is so unforgiving as to condemn people to eternal hell, then why should we not all feel like Bertrand Russell and do an about-face on Jesus? Like Russell, any modern Christian may feel this way.

The author Mark Mason (1997) has dealt with this subject quite well, and I refer the reader to his book. Mason demonstrates that Jesus never used the word "hell," nor did he mean it. It is because of errors in translation from the original Greek and the manipulations of the medieval Christian church that Jesus' image has been tarnished in this way. Mason also argues cogently that the word "forgiven" in connection with speaking against the Holy Ghost is an unfortunate error in translation as well, not in keeping with the context.

As to being unforgiving, many stories, such as the parable of the Good Samaritan (Luke 10: 29-37), suggest otherwise when properly analyzed (Mason, 1997). And who doesn't know about the episode when he protected a woman from being stoned to death by saying, "Let him who has not sinned throw the first stone"?

WAS JESUS AN AVATARA?

There is something else that a modern Christian may find interesting to consider. Hindus accept Jesus as an *avatara*, which is their word for a person who is fully transformed. It is believed that *avataras* incarnate as human beings whenever the movement of consciousness stagnates (whenever conscious evolution is stalled). Does this fit with Jesus' situation?

It does indeed. Hindus consider people like Krishna, Buddha, Shankara, and Ramakrishna as *avataras,* because they all came at a time when religion and spirituality stopped being a force in people's life. These *avataras* restored spirituality to their societies. Similarly, Jesus came to rescue Judaism from an intense period of stagnation.

Another parallel is well known. Krishna says in the *Bhagavad Gita*, "I am the goal of the wise and I am the way." As similarly Jesus said, "I am the way and the truth and the life."

And of course, Jesus did say:

I have other sheep that are not of this fold. I must bring them also. They too will hear my voice, and there shall be one flock and one shepherd. (John 10:16)

This resonates well with Krishna's declaration in the *Bhagavad Gita*:

In every age I come back
To deliver the holy,
To destroy the sin of the sinner,
To establish righteousness.

Indeed the parallels are striking. So again, was Jesus an *avatara*? Is the concept of *avatara* even acceptable to the new science of God and spirituality that we have built here?

I have argued elsewhere (Goswami, 2001) that people who are completely transformed (another word is "liberated") complete the death-birth-rebirth cycle. We can ask, what happens to their quantum monad with its fully perfected patterns of living when they die? Scientifically, we must concede, the quantum monad should be there in potentia available for future use.

Future use? How?

One use is for us to invoke such a quantum monad as a personal spirit guide through something like channeling. This we do. A Hindu has the option to use Krishna or Shankara as his or her spirit guide. Similarly, a Buddhist has Buddha, a Jew has Moses, a Moslem has Muhammad, and a Christian has Jesus.

A second use is to serve the needs of the evolution of consciousness. Whenever evolution stagnates, the evolutionary pressure brings about a rebirth of the quantum monad of the previous *avatara*. This is why Jesus says, "Before Abraham I am." An *avatara* does not accrue any karma during his life. He takes birth with the perfected conditioning of the same perfected quantum monad of the previous *avatara*.

All right, there you have it. If what I have presented here helps you orient yourself better as a Christian to the new integrative science, then do consider Jesus' words, "There shall be one flock and one shepherd." Obviously Jesus foresaw some sort of integration of all religions. Can the new science be the locus for a unifying dialogue among all the religions of the world? It is up to you to make that happen.

Bibliography

Adler, A. (1938). *Social Interest: Challenge to Mankind*. London: Faber & Faber.

Amabile, T. (1990). "Within you, without you: The social psychology of creativity and beyond." In M. A. Runco and R. S. Albert (Eds.), *Theories of Creativity*. Newbury Park, CA: Sage.

Ager, D. (1981). "The nature of fossil record." *Proceedings of the Geological Association*, vol. 87, 131-159.

Aspect, A., Dalibar, J. and Roger, G. (1982). "Experimental test of Bell's inequalities with time varying analyzers." *Physical Review Letters*, vol. 49, 1804-1806.

Aurobindo, S. (1996). *The Life Divine*. Pondicherry, India: Sri Aurobindo Ashram.

Aurobindo, S. (1955). *The Synthesis of Yoga*. Pondicherry, India: Sri Aurobindo Ashram.

Bache, C. M. (2000). *Dark Night, Early Dawn: Steps to a Deep Ecology of Mind*. NY: Paragon House

Banerji, R. B. (1994). "Beyond words." Preprint. Philadelphia, PA: St. Joseph's University.

Barasch, M. I. (1993). *The Healing Path*. NY: Tarcher/Putnam.

Barrow, J. D. and Tippler, F. J. (1986). *The Anthropic Cosmological Principle*. NY: Oxford Univ. Press.

Bass, L.(1971). "The mind of Wigner's friend." *Harmathena*, no.cxii. Dublin: Dublin University Press.

Bateson, G. (1980). *Mind and Nature*. NY: Bantam.

Bem, D. and Honorton, C. (1994). "Does psi exist? Replicable evidence for an anomalous process of information transfer." *Psychological Bulletin*, January issue.

Behe, M. J. (1996). *Darwin's Black Box*. NY: Simon & Schuster.

Blood, C. (1993). "On the Relation of the Mathematics of Quantum Mechanics to the perceived physical universe and free will." Preprint. Camden, NJ: Rutgers University.

Blood, C. (2001). *Science, Sense, and Soul*. Los Angeles, CA: Renaissance Books.

Bohm, D. (1951). *Quantum Theory*. Englewood Cliffs, NJ: Prentice Hall.

Bohm, D. (1980). *Wholeness and Implicate Order*. London: Rutledge & Kegan Paul.

Briggs, J. (1990). *Fire in the Crucible*. Los Angeles, CA: Tarcher.

Byrd, C. (1988). "Positive and therapeutic effects of intercessor prayer in a coronary care unit population." *Southern Medical Jjournal*, vol. 81, 826-829.

Cairns, J., Overbaugh, J., and Miller, J. H., "The origin of mutants." *Nature*, vol. 335, 142-145.

Chalmers, D. (1995). *Toward a Theory of Consciousness*. Cambridge, MA: MIT Press.

Chopra, D. (1990). *Quantum Healing*. NY:Bantam- Doubleday.

Chopra, D. (1993). *Ageless Body, Timeless Mind*. London: Random House.

Cranston, S. L. and Carey, W. (1984). *Reincarnation*. 2 volumes. Pasadena, CA: Theosophical University Press.

Crick, F. and Mitchison, G. (1986). "The function of dream sleep." *Nature*, vol. 304, 111-114.

Csikszentmihalyi, M. (1990). *Flow: the Psychology of Optimal Experience*. NY: Harper Collins.

Cumming, H. and Leffler, K. (2006). *John of God: Healing through Love*. Hillsboro, OR: Beyond Words Publishing.

Darwin, C. (1859). *On the Origin of Species by Means of Natural Selection or the Preservation of Favored Races in the Struggle for Life*. London: Murray.

Davies, P. (1988). *The Cosmic Blueprint*. NY: Simon & Schuster.

Dawkins, R. (1976). *The Selfish Gene*. NY: Oxford University Press.

Dawkins, R. (2006). *The God Delusion*. Boston: Houghton Mifflin

Devall, W. and Sessions, G. (1985). *Deep Ecology*. Salt Lake City: Peregrin Smith.

Dossey, L. (1992). *Meaning and Medicine*. NY: Bantam.

Dusek et al. (2004). *American Heart Journal*, April 4 issue,

Einstein, A., Podolsky, B., and Rosen, N. (1935). "Can quantum mechanical description of physical reality be considered complete?" *Physical Review Letters*, vol. 47, 777-80.

Elredge, N. and Gould, S. J. (1972). "Punctuated equilibria: An alternative to phyletic gradualism." In *Models of Paleontology*, T. J. M. Schopf, ed. San Francisco, CA: Freeman.

Elsasser, W. M. (1981). "Principles of a new biological theory: a summary. *Journal of Theoretical Biology*, vol. 89. 131-50.

Elsasser, W. M. (1982). "The other side of molecular biology." *Journal of Theoretical Biology*, vol. 96, 67-76.

Feynman, R. P., Leighton, R. B., and Sands, M. (1962). *The Feynman Lectures in Physics*, vol. 1. Reading, MA: Addison-Wesley.

Goleman, D. (1994). *Emotional Intelligence*. NY: Bantam.

Goswami, A. (1989). "The idealist interpretation of quantum mechanics." *Physics Essays*, vol. 2, 385-400.

Goswami, A. (1993). *The Self-Aware Universe: How Consciousness Creates the Material World*. NY: Tarcher/Putnam.

Goswami, A. (1994). *Science within Consciousness*. Research Report. Sausalito, CA: Institute of Noetic Sciences.

Goswami, A. (1997a). "Consciousness and biological order: toward a quantum theory of life and evolution." *Integrative Physiological and Behavioral Science*, vol. 32, 75-89.

Goswami, A. (1997b). "A quantum explanation of Sheldrake's morphic resonance." In H. P. Durr and F. T. Gottwald, ed. *Scientists Discuss Sheldrake's Theory about Morphogenetic Fields*, Germany: Scherzverlag.

Goswami, A. (1999). *Quantum Creativity*. Cresskill, NJ: Hampton Press.

Goswami, A. (2000). *The Visionary Window: A Quantum Physicist's Guide to Enlightenment*. Wheaton, IL: Quest Books.

Goswami, A. (2001). *Physics of the Soul*. Charlottesville, VA: Hampton Roads.

Goswami, A. (2002). *The Physicists' View of Nature*, vol. 2. NY: Kluwere Academic/Plenum.

Goswami, A. (2003). *Quantum Mechanics*. Long Grove, IL: Waveland Press.

Goswami, A. (2004). *The Quantum Doctor*. Charlottesville, VA: Hampton Roads.

Goswami, A. (2008). *Creative Evolution: How Evolution Proves Intelligent Design*. Wheaton, IL: Theosophical Publishing House.

Grinberg-Zylberbaum, J., Delaflor, M., Attie, L., and Goswami, A. (1994). "Einstein Podolsky Rosen paradox in the human brain: the transferred potential." *Physics Essays*, vol. 7, p. 422-428.

Grof, S. (1998). *The Cosmic Game: Explorations of the Frontiers of Human Consciousness*. Albany, NY: SUNY Press.

Hadamard, J. (1939). *The Psychology of Invention in the Mathematical Field*. Princeton, NJ: Princeton University Press.

Harman, W. and Reingold, H. (1984). *Higher Creativity*. Los Angeles, CA: Tarcher.

Hellmuth, T., Zajonc, A. G., and Walther, H. (1986). In *New Techniques and Ideas in Quantum Measurement Theory*, ed. D. M. Greenberger. NY: N. Y. Academy of Science.

Hillman, J. (1992). *The Thought of the Heart and the Soul of the World*. Woodstock, CT: Spring Publications.

Hobson, J. A. (1990). "Dreams and the brain." In Krippner, S. (Ed.). *Dreamtime and Dreamwork*. NY: Tarcher/Perigee.

Hofstadter, D. R. (1980). *Goedel, Escher, Bach: An Eternal Golden braid*. NY: Basic Books.

Holmes, E. (1938). *Science of Mind*. NY: Tarcher/Putnam.

Humphrey, N. (1972). "Seeing and nothingness." *New Scientist*, vol. 53, p. 682.

Jahn, R. (1982). "The persistent paradox of psychic phenomena: An engineering perspective." *Proceedings of the IEEE*, vol.70, 135-170. NY: Carrol & Graf.

Jung, C. G. (1971). *The Portable Jung*, ed. J. Campbell. NY: Viking.

Krishnamurthy, U. (2008). *Yoga Psychology*. To be published.

Labarge, S. (1985). *Lucid Dreaming*. NY: Ballantine.

Laszlo, E. (2004). *Science and the Akashic Field*. Rochester, VT: Inner Traditions.

Lewontin, R. (2000). *The Triple Helix*. Cambridge, MA: Harvard Univ. Press.

Libet, B. (1985). "Unconscious cerebral initiative and the role of conscious will in voluntary action." *Behavioral and Brain Science*, vol. 8, 529-566.

Libet, B., Wright, E., Feinstein, B., and Pearl, D. (1979). "Subjective referral of the timing of a cognitive sensory experience." *Brain*, vol. 102, p. 193.

Liu, Y., K. Vian, P. Kckman. (1988). *The Essential Book of Traditional Chinese Medicine*. NY: Columbia University Press.

Lovelock, J. (1982). *Gaia: A New Look at Life on Earth*. Oxford: Oxford University Press.

Magallon, L. L., and Shor, B. (1990). "Shared dreaming: joining together in dreamtime." In Krippner, S. (Ed.). *Dreamtime and Dreamwork*. NY: Tarcher/Perigee.

Maslow, A. H. (1968). *Toward a Psychology of Being*. NY: Van Nostrand Reinhold.

Maslow, A. H. (1971). *The Further Reaches of Human Nature*. NY: Viking.

Mason, M. (1997). *In Search of the Loving God*. Eugene, OR: Dwapara Press.

Maturana, H. (1970). "Biology of cognition." Reprinted in Maturana, H. and Varela, F. (1980). *Autopoiesis and Cognition*. Dordrecht, Holland: D. Reidel.

Mergulis, L. (1993). "The debates continue." In Barlow, C. (Ed.) *From Gaia to Selfish Gene*, 235-238, Cambridge, MA: The MIT Press.

Merrell-Wolff, F. (1995). *Philosophy of Experience*. Albany, NY: SUNY Press.

Mitchell, M. and Goswami, A. (1992). "Quantum mechanics for observer systems." *Physics Essays*, vol. 5, 525-529.

Moss, R. (1981). *The I That Is We*. Berkeley, CA: Celestial Arts.

Moss, R. (1984). *Radical Aliveness*. Berkeley, CA: Celestial Arts.

Moura, G. and Don, N. (1996). "Spirit possession, Ayahuaska users and UFO experiences: three different patterns of states of consciousness in Brazil." Abstracts of talks at the 15th International Transpersonal Association conference. Manaus, Brazil. Mill Valley, CA: International Transpersonal Association.

Newberg, A. D'Aquili, E., and Rause, V. (2001). *Why God Won't Go Away*. NY: Ballantine.

O'Regan, B. (1987). *Spontaneous Remission: Studies of Self-Healing*. Sausalito, CA: Institute of Noetic Sciences.

O'Regan, B. (1997). "Healing, remission, and miracle cures." In Schlitz and Lewis (Eds.), *The Spontaneous Remission Resource Packet*. Sausalito, CA: Institute of Noetic Sciences.

Page, C. (1992). *Frontiers of Health*. Saffron Walden, UK: The C.W. Daniel Co. Ltd.

Peat, F. D. (1987). *Synchronicity*. NY: Bantam.

Perls, F. (1969). *Gestalt Therapy Verbatim*. Moab, UT: Real People's Press.

Penrose, R. (1989). *The Emperor's New Mind*. Oxford: Oxford University Press.

Pert, C. (1997). *Molecules of Emotion*. NY: Scribner.

Piaget, J. (1977). *The Developmenment of Thought: Equilibration of Cognitive Structures*. NY: Viking.

Radin, D. (1997). *The Conscious Universe*. NY: HarperEdge.

Radin, D. (2006). *Entangled Minds*. NY: Paraview Pocket Books.

Ring, K. (1984). *Heading toward Omega*. NY: William Morrow.

Ring, K. and Cooper, S. (1995). "Can the blind ever see? A study of apparent vision during near-death and out-of-body experiences." Preprint. Storrs, CT: Univ. of Connecticut.

Sabel, A., Clarke, C., and Fenwick, P. (2001). "Intersubject EEG correlations at a distance—the transferred potential." In Alvarado, C. S. (Ed.) *Proc. of the 46th Annual Convention of the Parapsychological Association*, 419-422.

Sabom, M. (1982). *Recollections of Death: A Medical Investigation*. NY: Harper & Row.

Schlitz, M. and Grouber, E. (1980). "Transcontinental remote viewing." *Journal of Parapsychology*, vol. 44, 305-317.

Schlitz, M. and Honorton, C. (1992). "Ganzfieldpsi performance within an artistically gifted population." *Journal ASPR*, volume 86, 83-98.

Schlitz, M. and Lewis, N. (1997). *The Spontaneous Remission Resource Packet*. Sausalito, CA: Institute of Noetic Sciences.

Schmidt, H. (1993). "Observation of a psychokinetic effect under highly controlled conditions." *Journal of Parapsychology*, vol. 57, 351-372.

Schuon, F. (1984). *The Transcendent Unity of Religions*. Wheaton, IL: Theosophical Publishing House.

Searle, J. (1987). "Minds and brains without programs." In C. Blackmore and S. Greenfield (Eds.), *Mind Waves*. Oxford: Basil Blackwell.

Searle, J. R. (1994). *The Rediscovery of the Mind*. Cambridge, MA: The MIT Press.

Sheldrake, R. (1981). *A New Science of Life*. Los Angeles, CA: Tarcher.

Shapiro, R. (1986). *Origins: A Skeptic's Guide to the Creation of Life on Earth*. NY: Summit Books.

Sicher, F., Targ, E., Moore, D., Smith, H. S. (1998). "A randomized double-blind study of the effect of distant healing in a population with advanced AIDS—report of a small scale study." *Western Journal of Medicine*, vol, 169, 356-363.

Sivananda, S. (1987). *Vedanta (Jnana Yoga)*. Rishikesh, India: Divine Life Society.

Squires, E. J. (1987). "A viewer's interpretation of quantum mechanics." *European Journal of Physics*.

Standish, L. J., Kozak, L., Clark Johnson, L., and Richards, T. (2004). "Electroencephalographic evidence of correlated event-related signals between the brains of spatially and sensory isolated human subjects." *The Journal of Alternative and Complementary Medicine*, vol. 10. 307-314.

Stapp, H. P. (1993). *Mind, Matter, and Quantum Mechanics*. NY: Springer.

Stevenson, I. (1974). *Twenty Cases Suggestive of Reincarnation*. Charlottesville, VA: The University Press of Virginia.

Stevenson, I. (1977). "Research into the evidence of man's survival after death." *Journal of Nervous and Mental Disease*, vol. 165, 153-183.

Stevenson, I. (1987). *Children Who Remember Previous Lives: A Question of Reincarnation*. Charlottesville, VA: The University Press of Virginia.

Tagore, R. (1931). *The Religion of Man*. NY: Macmillan.

Taimni, I. K. (1961). *The Science of Yoga*. Wheaton, IL: Theosophical Publishing House.

Targ, R. and Katra, J. (1998). *Miracles of Mind*. Novato, CA: New World Library.

Targ, R. and Puthoff, H. (1974). ""Information transmission under conditions of sensory shielding." *Nature*, vol. 252, 602-607.

Teasdale, W. (1999). *The Mystic Heart*. Novato, CA: New World Library.

Teilhard de Chardin, P. (1961). *The Phenomenon of Man*. NY: Harper & Row.

Thom, R. (1975). *Structural Stability and Morphogenesis*. Reading, MA: Benjamin.

Tiller, W. A., Dibble, W. E., and Kohane, M. J. (2001). *Conscious Acts of Creation*. Walnut Creek, CA: Pavior Publishing.

Van Lommel, P., van Wees, R., Meyers, V., Elfferich, I. (2001). "Near-death experiences in survivors of cardiac arrest." *The Lancet*, vol. 358, 2039-2045.

Visser, F. (2003). *Ken Wilber: Thought as Passion*. Albany, NY: SUNY Press.

Vithulkas, G. (1980). *The Science of Homeopathy*. NY: Grove Press.

Von Neumann, J. (1955). *The Mathematical Foundations of Quantum Mechanics*. Princeton: Princeton University Press.

Von Neumann, J. (1966). *The Theory of Self-Reproducing Automata*. Urbana, IL: University of Illinois Press.

Wackermann, J., Seiter, C., and Holger, K. (2003). "Correlation between brain electrical activities og two spatially separated human subjects." *Neuroscience Letters*. vol. 336, 60-64.

Waddington, C. (1957). *The Strategy of the Genes*. London: Allen and Unwin.

Wallas, G. (1926). *The Heart of Thought*. NY: Harcourt, Brace & World.

Wambach, H. (1978). *Reliving Past Lives: The Evidence under Hypnosis*. NY: Harper & Row.

Weil, A. (1983). *Health and Healing*. Boston: Houghton Mifflin.

Weil, A. (1995). *Spontaneous Healing*. NY: Knoff.

Wickramsekera et al. (1997). "On the psychophysiology of the Ramtha School of enlightenment." Preprint.

Wilber, K. (1981). *Up from Eden*. Garden City, NY: Anchor/ Doubleday.

Wilber. K. (2000). *Integral Psychology*. Boston: Shambhala.

Wolf, F. A. (1970). *Space, Time, and Motion*.

Woolger, R. (1988). *Other Lives, Other Selves*. NY: Doubleday.

Index

About the Author

 Amit Goswami is a theoretical nuclear physicist. Goswami received his PhD in physics from the University of Calcutta in 1964 and moved to the United States early in his career. He taught physics for 32 years as a member of The University of Oregon Institute for Theoretical Physics. Starting at age 38, his research interests shifted to quantum cosmology, quantum measurement theory, and applications of quantum mechanics to the mind-body problem. These days he is probably best known as one of the interviewed scientists featured in the 2004 film *What the Bleep Do We Know!?* He is also featured in the recent documentary *Dalai Lama Renaissance* and is the subject of the 2009 documentary *The Quantum Activist.*

Fully retired as a faculty member since 2003, Dr. Goswami now speaks nationally and internationally. He is also a member of the advisory board of the Institute of Noetic Sciences, where he was a senior scholar in residence from 1998 to 2000.

Dr. Goswami is the author of *How Quantum Activism Can Save Civilization, Physics of the Soul,* and *The Quantum Doctor.* Find out more about Dr. Goswami at *amitgoswami.org*

Hampton Roads Publishing Company

. . . for the evolving human spirit

HAMPTON ROADS PUBLISHING COMPANY
publishes books on a variety of subjects,
including spirituality, health, and other
related topics.

For a copy of our latest trade catalog,
call 978-465-0504 or
visit our website at www.hrpub.com